D1711197

MANAGING CORPORATE MEDIA

By Eugene Marlow, Ph.D.

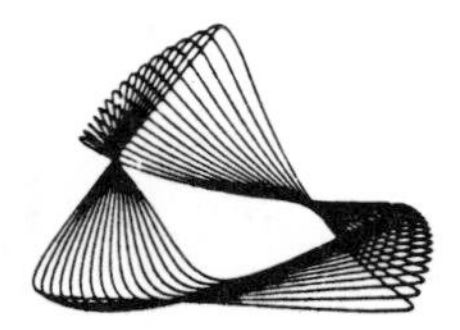

Knowledge Industry Publications, Inc.
White Plains, NY

The Video Bookshelf

Managing Corporate Media

Library of Congress Cataloging-in-Publication Data

Marlow, Eugene.
Managing corporate media.

(The Video bookshelf)
Rev. ed. of: Managing the corporate media center.
Includes index.
1. Business libraries—Administration. 2. Corporate libraries—Administration. 3. Instructional materials centers—Management. 4. Corporations—Information services—Management. 5. Business communication.
I. Marlow, Eugene. Managing the corporate media center.
II. Title. III. Series.
Z675.B8M36 1989 027.6'9 88-27179
ISBN 0-86729-265-2

Printed in the United States of America

10 9 8 7 6 5 4 3 2 1

To Judy, Jonathan and Samuel

Table of Contents

Tables and Figures

Foreword

$13.1 billion! This is what Hope Reports (Rochester, NY) estimates business and industry, education, government, health and religious organizations spent on media in 1987. By "media," Hope Reports means the use of slides, overheads, filmstrips, video, multi-image, film, microcomputers and videoconferencing for internal and external communications. The $13.1 billion compares to $8.5 billion spent on media by organizations in 1981.

Corporate organizations alone spent $9 billion (est.) in 1987 compared to $5.4 billion spent in 1981—a 67% increase in six years.

Between 1975 and 1985, organizations spent more on slides than they did on video, and more on video than they did on film, as follows:

	1975	1985
Slides	$644 million	$5.3 billion
Video	$220 million	$1.6 billion
Film	$0.5 billion	$0.5 billion

Clearly, film's growth as a corporate communications medium has been flat since the middle of the last decade...and there is evidence that film's use in the corporate arena diminished significantly in the latter part of the 1980s. Further, corporate America's use of print dwarfs the use of other filmic and videographic media. Dianna Booher, author of *Cutting Paperwork in the Corporate Culture* (New York: Facts on File, 1986), estimates "Paperwork costs more than $100 billion each year in the United States."

In the aggregate, therefore, corporate America's use of various graphic, filmic and videographic media is significant from an economic perspective and highly important in terms of the communications functions these various media perform.

Managing Corporate Media is about managing media used for a variety of internal/external communications purposes in the corporate context. In this volume, graphic media is defined as print and graphics. Filmic media is defined as overheads, slides, still photography and 16mm film. Electronic media refers to audiotape, computer graphics, videotape, multi-image, videodisc and teleconferencing.

While most organizations engage in media production operations (using either internal resources, external resources or a combination of the two), many organizations do not centralize their media production operations. It is this author's opinion that there are inherent relationships among the various communications media and that, while many organizations evolve their media production functions in a hodge-podge manner, it does not have to be that way. The social and technological pressures of the 1980s and 1990s will preclude such an evolutionary process—which, in turn, will lead more organizations to initiate, evolve and develop their media production operations in a more rational fashion.

Since the first edition of this book was published (in 1981 as *Managing the Corporate Media Center*) several significant changes have taken place. First, there has been an increase in the number and kind of electronic devices used by corporate America. Hopefully, the new chapter on using computers and the revised chapter on the evolving media operation address these changes.

Second, the statistics clearly show that print and slides are still the leaders in terms of dollars spent by corporations for communications activities. It is also clear that video has grown as a corporate communications tool. But the economic realities of the 1980s and (presumably) the 1990s have changed the nature of so-called corporate video operations. A visible number have been shut down, many have been down-sized. A few have turned to external clients for business (see Chapter 8 on marketing). Apparently many corporations are using external resources more frequently to realize their video communications product.

Perhaps these changes reflect the changing landscape of corporate America: fewer larger corporations (as a result of mergers) and an increasing number of smaller companies (200 employees and less). Companies are also flattening the organizational chart, resulting in fewer layers of management between the top and the bottom. This may be a direct result of the increasing use of electronic communications media (such as the computer).

It does not appear likely these changes will abate soon. The evolving nature of the world's economic structure bodes an increase in competition from within and without the United States, and this will probably mean even greater pressure on all companies to be "lean and mean." In turn, large-scale internal corporate video departments may eventually be replaced by small-scale departments using a variety of external production resources.

There are parallels in the commercial television industry as more and more television stations use technology to down-size production crews, electronic news departments and management layers.

The point is: in the management of corporate media production functions there is a great need for flexibility in response to the ebb and flow of American corporate life.

The aim of this book is to assist managers and organizations with both the development and the working realities of corporate media. It begins by looking at how and why a corporate media production operation is created, and then devotes individual chapters to examining the department budget, organization and staffing. This is followed by a discussion of how to solve communications problems, and a guide to day-to-day operations. The chapter on marketing the operation (to top management as well as to clients) applies some of the same communications techniques to the media operation that are generally used by its clients. The chapter on working with executives, clients and content experts ought to be of interest to the entire staff. We then look at the proper use of external resources, and end by discussing trends that we feel will particularly affect the evolving media operation of tomorrow. There are two new chapters: one on the use of computers and another on contracts, copyrights and permissions.

The appendices provide several resource lists of readings in media and management, professional organizations and publications and media competitions.

* * *

There are various persons I would like to thank whose expertise and comments contributed to the development of the current volume. First thanks to Tom Hope, president of Hope Reports, for providing the statistical information that puts the current work in appropriate economic context.

Much appreciation to Ellen Lazer, senior editor at Knowledge Industry Publications, Inc., for providing this author with that rare opportunity to go back and revise the first edition of this volume and make clearer what was perhaps vague, give greater definition where it was needed and bring up-to-date (as much as is possible with a printed work) content that was outdated.

And last, to my best and worst supervisors at Citibank, Prudential Insurance and Union Carbide, to my production and engineering staffs at Prudential and Union Carbide and all my clients since I formed Media Enterprises Inc in 1982—my thanks for all the experience these relationships brought: without them this volume could not have been written.

Eugene Marlow
1989

1

Creating the Corporate Media Function

Social and technological developments have expanded the scope and changed the complexity of corporate communications. Today corporate managers are faced with a multitude of communications problems that were either not considered problems 40 years ago or were far simpler: communications with government and regulatory bodies; with top executives, board members, stockholders and employees; with dealers and customers (increasingly on an international scale); with critics; with the press; even with other managers at corporate headquarters or locations halfway around the world.

The trend today towards demassification or "pluralism" continues to fragment our one-time "mass" society, creating even more complex communications problems for corporate managers. In addition, the many technological innovations of the last 40 years have involved communications devices that create greater access to more information by diverse publics: videotape, cable television, satellites, videodiscs, personal computers, online databases—all communications technologies that allow for pluralistic communications environments.

These social and technological developments have had their effect on corporate communications activities. A cursory analysis of contemporary corporate America would show a vast array of employee and customer training activities; communications networks among far-flung and diverse corporate entities; ever-expanding techniques for employee communications; changing channels for marketing and sales communications; and new methods for communicating with political, press and public groups. Furthermore, there are books, journals, newsletters, associations, seminars and even academic degrees in communications subjects that were unimaginable 40 years ago!

THE EMERGING CORPORATE MEDIA FUNCTION

Paralleling this growth in the quantity and complexity of corporate communications activities has been the development of corporate media departments—organizational functions that generate various print, graphic, film and electronic products for the purpose of solving a variety of organizational communications problems. Today's corporate media functions have sprung from, and continue to evolve in response to, the internal and external communications needs of America's corporations. The important point is that corporate media departments do not just happen—they develop in response to needs.

The last 40 years have shown that the communications needs of America's corporations have changed and will continue to change, and that the solutions to communications problems will also change. This chapter will show that the creation and development of a corporate media department that is properly organized, staffed, equipped and financed begins in response to the communications needs of the organization. Analysis of communications needs and media use then translates into a level of demand for media production resources. For many organizations, this leads to the creation, development and/or expansion of a media department. (See Figure 1.1.)

Figure 1.1: Overview: Determining the Need for a Corporate Media Function

Definition of Internal & External Communications Needs

↓

Definition of Media Needs (Graphic, Photographic/Cinemagraphic, Electronic)

↓

Media Production Resources Management

↓

Initiation and/or Development of the Corporate Media Department

COMMUNICATIONS ACTIVITIES

There are two broad areas of organizational communications activity: internal communications and external communications. A specific breakdown of these communications activities is given in Table 1.1.

Table 1.1: Internal and External Communications Activities

INTERNAL

Training

- Role playing
- Basic sales training
- Dealer sales training/customer training (operations/service)
- Product training
- Technical training:
  - Machine operation
  - Maintenance and repair
  - Troubleshooting
  - Precision techniques
- New procedures
- Proficiency upgrading
- Safety training
- Operations improvement
- Management development

Manufacturing

- Quality control
- Testing
- Methods improvement
- Documentation

Research & Development

- Recording live tests
- Communicating with marketing

General Employee Communications

- Communications from headquarters to the field
- Communications from the field to headquarters
- Policy announcements
- Marketing developments
- Management conferences
- Annual stockholders meetings
- Orientation programs for new hires
- Company benefits
- Periodic news

EXTERNAL

Sales Promotion

- Product demonstrations
- New product introductions
- Explanation of services
- Sales trip reports
- Sales meetings
- Reports on competition

Marketing

- Point of purchase
- Trade shows
- Reports on new applications
- Reports on new markets
- Testing new advertisements/promotion ideas
- Communicating with research and development

Table 1.1: Internal and External Communications Activities (cont'd.)

External Communications
Recruiting new hires
Community relations
Government relations
Security analyst relations
Press conferences
Press relations
General public relations
Consumer relations
Visitor orientation programs
Visual facilities tours
Convention exhibits

DEFINING COMMUNICATIONS NEEDS

Organizations of any complexity are fraught with these internal and external communications activities and problems, many of which are served and solved by the appropriate use of various media. The initial problem for the organization as a whole is to uncover existing or potential communications needs and translate them into a level of demand for media production services.

Background Research

How and where can information about communications needs be obtained? The organization's annual report is a good place to start. It can provide an overview of the company's activities and growth patterns. Other source documents include organizational charts that provide information about reporting relationships and recent press releases which will indicate major current organizational events.

The next step is to focus on divisions of the organization. Where are communications problems more likely to occur? Below we list a number of functions commonly found in most organizations. These are suggested starting points for investigation in terms of their levels of communication activity. Both background study and personal contact are recommended.

Organizational executives, such as the chairman of the board, the president, executive vice presidents and senior vice presidents.

Functional areas, such as accounting, auditing, building services, community affairs, computer services, corporate contributions, controller's group, corporate accounting, corporate long-range planning, corporate communications, employee relations, energy supplies and services, general services, government relations, health, safety and environmental affairs, human resources development, industrial relations, information systems planning, international affairs, investor relations, law, management development, medical, marketing services, office and information services, public affairs, purchasing, research and development, sales and marketing, secretarial, security, telecommunications and university relations.

Operations management, such as division president, division vice presidents, product managers, national sales managers, regional sales managers, production managers, administrative systems managers, quality control managers, operations managers, plant managers, office managers and customer services managers.

Conducting Interviews

Fact-finding interviews can be immensely helpful in learning about the communications needs, activities and problems of the organization and its parts. Here are some typical questions one might ask during the interviews:

What are your most common communications activities?
What are your most common communications problems?
What medium or media do you currently use?
Which medium or media are the most dominant?
Do you find them effective?
How are they currently being produced?
Are there changes on the horizon that might necessitate changes in how and what you communicate?
Will these communications activities increase in their level of activity or decrease?
How are internal and external communications currently distributed?

Organizing the Communications Patterns

The more interviews conducted, the more likely the conclusions will be validated. By the twentieth or thirtieth interview, a pattern will begin to emerge. As common communications problems and needs begin to make themselves known, they should be given top priority. After the fortieth—yes, or even the fiftieth interview!—one should have a very good overview of both the organization's operations and the communications problems either common to the entire organization or specific to parts of it. Every organization will be different. Some will exhibit many common communications problems across the board; others will exhibit problems that are specific to only one or more parts of the organization.

Figure 1.2 is a suggested matrix by which one can organize and analyze the information gleaned from the interviews.

Analyzing the Data

Once the interviews are completed and first impressions have been reviewed, the next step is to cut the fat away from the meat. Some problems may seem trivial while more important ones may stand out if only because they have been referred to by many executives. Common, important issues might include the need for an overview of the organization's operations for orientation, community relations and recruiting. Or, it might be vital to concentrate on informing employees (especially middle managers) of laws or regulations requiring compliance (such as environmental issues, business ethics or Securities and Exchange Commission regulations).

The responses might also be grouped according to which level the communications problem affects in the organization. Some communications have a corporate-wide audience, while others involve only a small, select portion of one division. Many communications problems fall somewhere between these two poles.

In order to facilitate the analysis, internal communications activities can be ranked by audience in the following order:

- all employees throughout the organization
- all managers throughout the organization
- all hourly employees throughout the organization
- certain professionals throughout the organization (such as purchasing agents or quality control managers)
- certain specialists throughout the organization (such as keypunchers or word processing operators)

The same kind of structure can be applied to problems that concern only one division of the organization:

- all employees in the division
- all managers in the division

Figure 1.2: Communications Needs Analysis Matrix

Communications Need	Audience	Communications Frequency	Current Media	Means of Distribution
Internal communications				
Training				
Manufacturing				
R&D				
General employee communications				
Orientation				
Benefits				
External communications				
Sales promotion				
Marketing				
Recruiting				
Community relations				
Public relations				
Government relations				
Press relations				
Financial relations				

- all hourly employees in the division
- professionals in the division (such as purchasing agents or quality control managers)
- certain specialists in the division (such as keypunchers or word processing operators)

External communications activities can also be ranked by audience, such as:

- customers
- recruits
- community leaders
- local government leaders
- state government leaders
- federal government leaders
- international government leaders
- union leaders
- university and college leaders
- security analysts
- press corps
- other corporate executives
- stockholders
- professional associations
- trade groups
- consumer groups

Once the communications needs have been slotted into audience categories, one must then look at the immediacy of the problems: How many prospective clients need something done right away? How many can wait three to six months? How many do not need anything until next year? How many have a continuing need? The answers to these questions can then be put into three categories: short term, intermediate term and long term.

Short term refers to anything that must occur in a three-to-six month period; intermediate term refers to anything that must happen within a six-month to 18-month period. Long term refers to anything that must happen in a period over 18 months. Juxtaposing communications needs by content categories against the time factors should result in a solid overview of the organization's short- to longer-range communications needs.

DEFINING THE MEDIA NEEDS

Before ultimately translating these communications activities (needs) and problems into an organizational media department, one must decide which media will be required. The various media under consideration in this book (see also Chapter 3, "Organizing the Media Production Function") are:

- teleconferencing
- videodisc
- multi-image
- videotape
- audiotape
- film
- computer graphics
- slides
- still photography
- graphic arts
- overheads
- print

Characteristics of Communications Media

Each medium has particular attributes which must be understood in order to effectively decide which to use for various situations. Table 1.2 summarizes some of the primary characteristics of the various media outlined above.

Table 1.2: Primary Characteristics of Communications Media

Medium	Primary Characteristics
Teleconferencing	Excellent for live, face-to-face meetings when travel/lodging expenses plus wear and tear on executives exceed cost of hardware installation and use expense over a three-to-five-year period. Makes meetings more efficient and effective. Helps to increase managerial productivity. Can be used on an ad hoc basis or with permanantly installed systems.
Videodisc	In high volume, duplication is less expensive than for videotape. Random access capability allows for a high degree of interaction. Excellent for training, general information communications and even sales promotion.
Multi-image	Can show various aspects of a single object, many objects or various time frames simultaneously. Can use slides, film, videotape and audio, as well as "live" elements simultaneously. Excellent for large audience viewings. Can be highly dramatic; not easily transportable.
Videotape	Generally less expensive to produce and duplicate than film. Can use film, slides, multi-image, audio, still photography, graphic arts and print elements. Excellent for communicating content involving motor skills and for attitude/behavior modeling. Communicates personality very well and provides immediate, live communications impact. It has immediate playback capability and is much faster to edit than film. Proliferation of VHS format creating wider distribution channels.

Table 1.2: Primary Characteristics of Communications Media (cont'd.)

Medium	Primary Characteristics
Audiotape	Relatively inexpensive to produce and distribute compared to videotape, film or slides. Can communicate information while the listener is focusing on other activities such as reading or even driving. Indispensable to videodisc, videotape, film and multi-image production.
Film	Increasing film stock, editing and duplication costs make film less cost-effective, and therefore less viable. In certain cases, however, film is easier to use for production than videotape. Excellent for photoanimation. Good for large audience viewings, although the development of large-screen video projectors may end this advantage. Useful when video players are not available.
Computer graphics	Rapidly growing corporate communications medium. Entire presentations can be created in short periods of time. Allows for some special effects. Information can be changed almost instantly.
Slides	Relatively inexpensive to produce, although price per slide may range from $15 to $150. Easily replaced when information becomes obsolete. Communicates facts effectively.
Still photography	Relatively inexpensive to originate. Does not require crews, as do video, film, multi-image and videodisc productions. Relatively inexpensive capital investment required. Photography can be used in many other media, including teleconferencing, videodisc, multi-image, videotape, film, slides and print.
Graphic arts	Indispensable media production activity to virtually every other communications medium.
Overheads	Inexpensive and fast medium for presenting facts. Good for simple, low-budget projects.
Print	Enormous applicability. Used in conjunction with virtually every communications activity and medium. It is hard to imagine a communications activity without some print piece associated with it.

The dozen communications media just described have been listed in a hierarchical order depending on communications function: each of the media, starting with print at the end of the list, provides a base for the communications medium above it.

For example, a print piece can be used in a teleconference, but a teleconference cannot be part of a print piece. Thus,

- a print piece can become part of an overhead
- an overhead can become part of a graphic piece
- a graphic can become part of a still photograph
- a still photograph can become part of a slide
- a slide can become part of a film
- a film can become part of a videotape
- a videotape can become part of a multi-image presentation
- a multi-image presentation can become part of a videodisc
- and a videodisc can become part of a teleconference.

Naturally, this paradigm is not airtight. For example, because computer graphics devices are generated electronically, they can create their own content without dependence on another medium. But the paradigm does hold true considering that the computer graphic may be part of a film, but not vice versa.

It is also true that, for the most part, the balance of the hierarchy cannot be reversed. For example, a live teleconference cannot become part of a videodisc; although a videodisc could be included in a multi-image presentation. However, in this case, the videodisc would only serve as a video source: the audience would not be able to access the videodisc in an interactive mode.

Continuing down the media hierarchy, the elements of a multi-image presentation can be included in a videotape, but the elements would have to be reformatted to accommodate television's smaller screen. And, a videotape could become part of a film, but an electronic process would be necessary to accomplish this.

Choosing the Right Communications Media

There are various criteria for matching one medium (or media) with a communications activity:

- communications frequency
- audience size
- audience location
- communications content.

Regarding communications frequency, if the communications content changes from day to day, or even from quarter to quarter, then media that are flexible and have relatively quick production turnaround times should be used. On this basis, teleconferencing, audio and videotape, slides, computer graphics, overheads and print formats are most practical and effective. These media are effective for short-term communications activities primarily because of their ability to create the communications package quickly. Communications which will remain fairly unchanged over the long term may be served best by videodisc, film and certain print products.

Audience size is a second criterion for choosing appropriate communications media. For large audiences, film, multi-image, videotape (using projection systems) and slides are most effective. For small audiences, virtually any medium can be used.

The location of the audience is another important factor. In small offices, for example, where there are no videotape players, a print piece may have to do the job.

The content of a communication will also help determine which medium is the most effective. Videotape is an excellent medium for role playing. Film and videotape are best for demonstrating motor skills. Electronic and film media are most suitable when content is highly visual in nature. Print media are best when content is primarily cognitive or factual in nature. Print media also allow the receiver an opportunity to review, study and analyze the communications content at a level of convenience not readily available with teleconferencing, multi-image or film. (More detailed discussion of "communications content" is found in Chapter 5, "Solving Communications Problems.")

Based on the above, communications activities can benefit from selecting appropriate media by application (both individually and in conjunction with one another). Table 1.3 illustrates this further.

Table 1.3: Selecting Media by Application

Communications Activity	Potential Media
INTERNAL COMMUNICATIONS	
General Employee Communications	
Annual stockholders meetings	Teleconferencing, videotape, film, slides, multi-image, print
Communications from field locations to headquarters	Videotape, print
Communications from headquarters to field locations	Videotape, film, print
Company benefits	Videodisc, videotape, film, computer graphics, slides, print
Management conferences	Teleconferencing, videotape, film, slides, multi-image, computer graphics, overheads, print
Marketing developments	Videotape, slides, print
Medical information	Videodisc, videotape, film, slides, print
Orientation program for new hires	Videodisc, videotape, film, slides, print
Periodic news	Videotape, print
Policy announcements	Videotape, print

Table 1.3: Selecting Media by Application (cont'd.)

Communications Activity	Potential Media
Manufacturing	
Documentation	Videotape, film
Methods improvement	Videotape
Quality control	Videotape, print
Testing	Videotape
Research & Development	
Communicating to marketing	Teleconferencing, videotape, audiotape, slides, print
Recording of live tests	Videotape, film
Training	
Basic sales training	Videodisc, videotape, audiotape, film, slides, multi-image, print
Dealer sales training/customer training (operations/service)	Videodisc, videotape, film slides, print
Management development	Videodisc, videotape, audiotape, film, slides, print
New procedures	Videodisc, videotape, film, slides, print
Operations improvement	Videodisc, videotape, film, slides, print
Product training	Videodisc, videotape, film, slides, print
Proficiency upgrading	Videodisc, videotape, film, slides, print
Role playing	Videotape
Safety training	Videodisc, videotape, film, slides, print
Technical training:	
Machine operation	Videodisc, videotape, film, slides, print
Maintenance and repair	Videodisc, videotape, film, slides, print
Troubleshooting	Videodisc, videotape, film, slides, print
Precision techniques	Videodisc, videotape, film, slides, print
EXTERNAL COMMUNICATIONS	
General External Communications	
Community relations	Videotape, film, slides, print
Consumer relations	Videotape, film, slides, print
General public relations	Videotape, film, slides, print
Government relations	Videotape, film, slides, print
Press conferences	Videotape, film, slides, print
Press relations	Videotape, film, slides, print
Recruiting	Videodisc, videotape, film, slides, print
Security analyst relations	Videotape, film, slides, print

Table 1.3: Selecting Media by Application (cont'd.)

Communications Activity	Potential Media
Visitor orientation programs	Videotape, film, slides, multi-image, print
Visual facilities tours	Videotape, film, slides, multi-image
Marketing	
Communicating with research and development	Teleconferencing, videotape, audiotape, print
Point of purchase	Videodisc, videotape, film, print
Reports on new applications	Videotape, slides, print
Reports on new markets	Videotape, slides, print
Testing new advertisements/ promotion ideas	Videotape, film, print
Trade shows	Videodisc, videotape, film, slides, multi-image, print
Sales Promotion	
Explanation of services	Videodisc, videotape, film, print
New product introductions	Videodisc, videotape, film, print
Product demonstrations	Videodisc, videotape, film, print
Reports on competition	Videotape, audiotape, slides, print
Sales trip reports	Videotape, audiotape, print
Sales meetings	Videotape, film, slides, multi-image, print

You may be asking: how do I select which medium to use if each application can be served by various media? The answer is that, for the most part, no communications application uses one medium to the exclusion of others. For example, there is virtually no communications activity that does not in some way use a print piece associated with it, even if that print piece is as basic as a label. More important, because each medium carries certain kinds of information better than others, it is not surprising to find various media working in tandem.

At a conference on artificial intelligence held by Texas Instruments a few years ago, the company used live teleconferencing to reach audiences in various parts of the world simultaneously. At the various locations, the company provided audience members with print pieces to reinforce the live presentations. Moreover, as people left the meeting, they were given a videotape (in VHS format) as a follow-up presentation (together with additional information in print format).

Thus, the reader should keep in mind that media work together in complementary fashion. The media manager working with other managers in the organization must choose which media will best serve each communications application and in what ratio: primarily print, primarily video, primarily slides, and so on.

DEFINING MEDIA PRODUCTION RESOURCES

Thus far we have outlined how the organization can define its communications needs and given them a rank order in terms of (1) applicability to the organizational hierarchy, and (2) immediacy, and further, how those needs can be translated into the media required to support them. The next step is to uncover what resources—internal and external—are available to satisfy media production needs, and what will be needed in the future. In other words,

(1) Where are we now?
(2) Where do we want to go?
(3) How will we get there?

At this juncture the organization must translate into real terms media solutions to the organization's stated, perceived communications needs, be they training, marketing, sales, etc. These needs must then be discussed in terms of people, hardware, facilities and the operating budget. To accomplish this, each of the three questions asked above should be examined in more detail.

Assessing the Resources on Hand

Where are we now? What media production resources is the organization presently using to satisfy its current and immediate-future media production needs? Questions to ask include:

How many people are on staff to produce video programs? films? multi-image productions?
How many photographers are on staff?
How many graphic artists are on staff?
How many video engineers and/or technicians are on staff?
What equipment (hardware) do we have in-house for video production? for film production? for multi-image production?
What equipment does the organization have in-house for slide and chart making?
What equipment does the organization have in-house for graphics production?
What facilities (space) does the organization have in-house for video production? film production? multi-image production?
What facilities does the organization have in-house for photography (such as adequate plumbing, electrical outlets, etc.)?
What facilities does the organization have in-house for graphics design and execution?

What is the current budget for video production? film production? multi-image production? photography production? slides and chart making?

What use is the organization making of external resources in the areas of video, film, photography, slides, charting and graphics arts? Who are the outside vendors?

What is the quality level of current in-house media production services?

What is the quality level of current external resources with respect to media production services?

What is the perception of organization executives with regard to in-house media production services?

What is the perception of organization executives with regard to external media production resources?

How are current in-house media production resources organized? Are they in one department? Two or more departments?

What is the current availability of teleconferencing facilities, videodisc and videotape players, film projectors and slide projectors?

Figure 1.3 provides a suggested matrix for organizing the answers to these questions.

Figure 1.3: Where Are We Now?

	Graphic Arts	Film	Electronic Media
Personnel (number):			
Equipment (type and value):			
Facilities (space, special requirements):			
Expense budget:			
Current applications (internal/external communications):			
Level of quality:			
How and where organized:			
Distribution systems:			
External resources used:			

Determining Future Needs

Where do we want to go? This refers to where the organization sees itself in three to five years with respect to the use, organization, staffing, equipping and budgeting of its media production functions—video, film, slides, multi-image, photography, graphic arts, teleconferencing, etc.

The answers to this question should reflect an analysis of the responses gleaned from the executive interviews. The overriding concept at this stage is that all future recommendations for staffing, equipping and budgeting grow out of the kinds of programs slated for production and the media that would be most effective for producing them. Once the short-to-long-range software needs have been determined, these applications can be translated into people: for example, how many video producers will be needed in the next six to 12 months? in the next one to two years? in the next three to five years? The same can be asked for photographers, graphic artists, video engineers, slide and charting specialists, etc.

Based on the answers given one can begin to determine equipment needs: what type of video production and post-production equipment is necessary? How many still cameras are needed? How many cameras for slide production? What kind of lighting instruments? What type of audiotape and multi-image production equipment?

The amount of personnel and equipment necessary will in turn determine the facilities needed: How much space will be required? How much heating, ventilating and air conditioning (HVAC)? Any special requirements regarding acoustics, plumbing, electricity, ceiling height, storage, color, lighting, meeting rooms, screening rooms, library, workrooms, maintenance room? How many office spaces will be needed?

Further, what is the need for distribution equipment, such as videodisc and videotape players, film and slide projectors, teleconferencing facilities? What kinds of changes will be needed?

Plotting the Course of Action

How will we get there? This question is essentially one of timing, i.e., not if, but when. By now the organization has determined what action must be taken (e.g., hiring staff, purchasing equipment, etc.) in order to meet its long-term communications goals. What remains is to integrate the organization's long-term media production needs with the realistic short-to-intermediate term budget constraints and, perhaps even more so, with management's perception of what is important. Figure 1.4 suggests a way to organize this information (please also refer to Chapter 2, "Budgets").

DEVELOPING A STRATEGY

There are several reasons for having an internal media production operation; namely accessibility, geographical convenience and maintaining the security and confidentiality of proprietary information. An organi-

Figure 1.4: How Do We Get There?

	YEAR I	YEAR II	YEAR III	YEARS IV-V
Volume of production				
Graphic arts				
Still photography				
Slides				
Cinematography				
Audiotape				
Videotape				
Multi-image				
Videodisc				
Teleconferencing				
Number of personnel				
Graphic arts				
Film				
Electronic media				
Equipment budget				
Graphic arts				
Film				
Electronic media				
Facilities				
Graphic arts				
Film				
Electronic media				
Expense budget				
Graphic arts				
Film				
Electronic media				

zation may need weekly or even daily access to media production personnel and facilities. Second, the organization may be too far from a metropolitan center with readily available external resources. Finally, an organization may decide to provide for an in-house media production operation if only to control the security of proprietary information.

The Economics of Internal and External Media Production

Much of the "when" decision to initiate, expand, modify or even contract media production operations is based on volume, which is another way of saying economics. Take video as an example. If an organization produces 15 or fewer programs a year, it is probably economically better off using outside resources. If the company is producing 15 to 35 programs a year, it could consider buying production equipment, but using outside facilities for editing. Last, if the company is producing 35 to 50 or more programs a year, it may be more cost-effective to have both production and post-production capabilities in-house. (Of course, these figures are somewhat arbitrary; moreover, they are irrelevant if external facilities are not accessible.) Also it should be remembered that having an in-house facility does not exclude the use of external resources, the topic of Chapter 10.

On the other hand, it is vital for any organization to have some kind of print duplication operation in-house, especially for small jobs. Imagine a company that did not even have a modest photocopying machine! It would be akin to not having a telephone system.

The issue, then, is not whether one medium or another should or should not be used, but whether the level of use warrants the development of an in-house function.

A company does not necessarily save money merely by buying and owning equipment. A 1-inch videotape recorder or Forox slide camera does not cost the organization less than it costs an outside facility. It is through the people a company hires to use the equipment that savings can be achieved.

Generally speaking, the ratio of internal to external aggregate costs is frequently 1:2. In other words, what costs 50 cents to produce in-house often costs one dollar to produce externally. In certain instances, savings are achieved by owning the equipment (particularly AV equipment—16mm film and 35mm slide projectors—that may be loaned to users within other departments of the organization). When it comes to video, film, multi-image and photography production, however, it is clear that the real savings are created by using on-staff personnel, as opposed to contracting exclusively with outside personnel. Levels of professionalism notwithstanding, having people in-house to manage media production can reduce production expenses from 25% to 50%, as compared to turning over the entire job to an outside company. However, the economics will vary from region to region and company to company.

The development of an in-house media operation, therefore, could well occur over a period of years. In the short term, a strategy may call for the initial hiring of a media manager and, later, in-house producer/directors who use outside production (hardware) facilities. Over the next one-to-two-year period, as volume grows, the economics may demand the purchase of equipment and hiring of technicians. If the volume of production grows, the purchase of additional equipment may be warranted. Increased volume may also herald the need for additional producer/directors and administrative support personnel.

The central factor here is volume: one to three films a year does not warrant the development of an in-house cinemagraphic unit. Twelve to

15 video programs a year does not warrant the creation of an in-house group (except under extreme circumstances). The need for a photographer or graphic artist two to three days a month does not support the need for a full-time, in-house photographic or graphic arts operation.

A way of looking at the "economic timetable" factor is to take each aspect of the media production operation and compare its internal cost to its external cost. One can then judge when it becomes economical to hire personnel, purchase equipment and/or expand the operating expense budget.

Let's look at two examples. A staff photographer is hired at an annual salary of $30,000 (including benefits). If an outside photographer charges $500 per day plus expenses (a low figure in some parts of the country), then the staff photographer is highly cost-effective. Presuming there are 250 working days in the year and the organization has enough volume of work for a photographer every working day, an outside photographer would cost the organization $125,000 a year! Conversely, if an organization needs a photographer only 50 days out of the year, then an outside photographer would prove more cost-effective (50 days x $500 = $25,000 per year).

On the equipment side, presume an organization is considering the purchase of a $50,000 video camera. Figure it costs this particular organization $500 a day to rent a video camera. If this organization is producing 30 programs a year, the economic payback of purchasing the camera is 3.3 years. Purchase of the camera, of course, may lead to the increased use of video as a communications medium, which would result in a faster payback on the capital investment.

In effect, the organization must make an economic judgment about (1) staff requirements, (2) capital equipment requirements and (3) concomitant operating budget expenses for each communications medium perceived as needed by the organization. Again, the overriding criterion is volume. When the production activity gets high enough (depending on the medium in question and relative external costs), it is time to hire personnel, purchase equipment, expand the operating budget or do all three simultaneously.

One other issue remains: the quality of the people and equipment required. Two principles have proven most practical, effective and efficient: good people produce quality work, and the proper equipment does the right job.

The Process

Presuming it takes an average of four or five years from initiation of an in-house media production operation to maturity, the step-by-step process might resemble the one illustrated in Table 1.4.

For an organization that has an existing media production operation, a year-to-year analysis might provide evidence that a reorganization of media production functions is warranted, or that some part of the operation should be expanded, or others contracted. Such analyses should be performed routinely by media production management, as well as by higher management. Figures 1.3 and 1.4, referred to earlier, provide suggested matrices to use for these types of analyses.

In sum, the order of events in the process is (1) analysis, (2) short- and long-range plans, (3) hiring of people, (4) purchase of equipment and, as volume grows, (5) the appropriate proportions of additional equipment and personnel—in that order!

Table 1.4: Chronological Development of a Media Production Operation

Year I	Needs analysis—performed either solely by in-house personnel or in conjunction with outside consultants familiar with communications needs analysis techniques, media production and long-range planning.
	Develop short- and long-range plans.
	Organization designates individual to coordinate media production operations.
	Organization makes extensive use of external resources.
Year II	As volume increases, organization hires production personnel: graphic artists, photographers, video/film/ multi-image producers, as required.
	Essential hardware purchased. External resources continue to be used (unless unavailable for geographical reasons).
Year III	As volume continues to grow, more hardware purchased and additional staff hired, particularly technical personnel.
Years IV to V	Additional software personnel hired. External resources continue to be used.

THE PLAN

Funds. Money. Capital. Dollars. Eventually, all the research, analysis, structuring, looking at the present, the future and ways of getting to the future boil down to a dollar figure: a year-by-year budget projection. (See

Chapter 2 for a more detailed discussion of budgets.) Let us assume for the moment, however, that budgets have been developed. The final step in this evaluation process is the preparation of a plan.

First, plans should be direct and clear. Second, they should be succinct. Third, they should be thorough. Fourth, they should be cogent.

The plan should have several content elements. The opening should contain a short introduction to the scope and purpose of the plan, sometimes called an "Executive Summary." The next section should follow the structure discussed earlier, namely: where are we now? where do we want to go? how do we get there? This section should be an explication of the current state of media production affairs, followed by a projection of where the organization should be in three to five years, concluding with how the organization should develop media production resources in that time period.

Other sections of the proposal could include a survey on the use of media in other organizations, especially those within the organization's industry. One might also include a chart showing cost comparisons between maintaining the use of outside resources, for example, versus developing an in-house resource (with thoroughly researched outside costs). Such a chart should cover at least a five-year period. Items such as software costs (for producer/directors, photographers, graphic artists, writers) should be shown, along with costs for hardware purchased. Maintenance as well as overhead costs should also be included in the cost analysis. Projected production volume growth curves should also be indicated; that is, at what point does the cost of in-house production become more economical (as well as convenient) to the organization, than the exclusive use of outside resources?

Finally, one might also include, in an appendix, the list of organizational executives interviewed and the analysis of the primary communications problems articulated. It is, after all, this information that has led to the conclusions that form the basis of the plan.

The quantification of facts based on thorough research presented in a succinct manner is more helpful than a barrage of words. Organizational media production is not entertainment but business, and must reflect a direct relationship between use and economic impact. For an organizational media production operation to be successful, it must accomplish the following task: the effective and efficient production of media that communicates information effectively and efficiently.

The success of an organizational media production operation, from an economic point of view, is based on developing a cost-effective in-house media operation and producing media products that provide positive economic results for the organization.

2

Budgets

At the heart of a successful media production department is a well-planned, well-coordinated and well-controlled budget. Without one, the organizational media function is a mere appendage on the side of the organizational body without a defined function or realizable benefit. With a properly structured annual budget, the media department can flourish and provide management with proof that it is making a contribution to the organization.

The development of an annual budget must be a direct outgrowth of the needs analysis (discussed in Chapter 1). In this sense, the development of the budget is a matter of economic analysis—frequently defined as the optimal use of scarce resources to satisfy human wants. In the case of media operations, human wants include the need to communicate information effectively and efficiently. Scarce resources include capital and operating dollars to staff, equip and provide operating monies to support communications activities. The optimal use of these scarce resources is what this book is about.

THE BUDGET'S FUNCTION

The budget can be defined as having three functions: (1) planning future operations; (2) coordinating the organization's activities; and (3) controlling the action of employees.[1] For the corporate media department, a good budget system provides planning by determining the demand for media production services; the expected cost of providing those services;

the amount and cost of equipment required; and the operating expenses required. Control of the operation is achieved by means of periodic reports (either monthly, quarterly, or semi-annually) for each part of the media operation. These budgetary performance reports reflect the budget amounts for each activity against the real expenditures.

Planning Future Operations

A periodic review of these budgets gives management an opportunity to investigate major differences between budget and actual performance. It also allows management to ascertain why predetermined plans have or have not been met and, in the latter case, to consider alternative courses of action. The budget may be thought of as a formal, written statement of management's plans for the future, expressed in financial terms. A budget charts the course of future action. It serves management in the same manner the architect's blueprints assist the builder. Like a blueprint, a budget should contain attainable objectives rather than mere wishful thinking.

A budget encourages planning because careful study, investigations and research must be given to expected future operations if the budget is to contain sound, viable goals. Advanced planning, in turn, increases the reliance of management on fact-finding in making decisions and lessens the role of hunches and intuition in managing the operation.

Organizational Coordination and Control

Coordination is facilitated as each level of management in the media operation participates in the preparation of the budget. In addition, a budget enables higher management to explain its objectives to each stratum in the organization and to keep these goals before the entire organization. As a result, employees within the operation are more integrated. Budgeting also contributes to effective control through the preparation of periodic budget reports. Finally, budget objectives also serve to encourage effectiveness and efficiency, cost savings, and have the potential for serving as a deterrent against waste.[2]

THE PARTS OF A BUDGET

The model for developing a media budget is no different than one for any other aspect of an organization. Speaking in marketing terms, the manager of the in-house media operation should adopt the attitude that his or her operation is an in-house "agency," in competition with any other "media agency" in the organization and any and all externally based media production operations. Thus the need for a proper, business like budget becomes even more obvious. As we proceed through the various steps and aspects of developing a budget, keep in mind that this budget is a business budget.

For planning, coordinating and control purposes, we must really deal with two types of budgets: an operating budget and a capital budget. The operating budget refers to expenses incurred during the fiscal year. The capital budget, on the other hand, refers to the purchase of equipment or facilities usually of value over $500. The primary reason for separating these two kinds of budgets is that operating expenses will usually occur and be paid for in that year. Capital expenditures, however, refer to things like buildings and equipment—items which have a life of more than one year. Obviously, videotape equipment, cameras, lighting equipment, animation stands, audio recording equipment and photographic processing equipment have a life of more than one year and therefore should be considered separately from the operating budget. Use Figure 2.1 as a suggested format for developing the annual operating expense budget, and Figure 2.2 for the capital equipment budget.

Figure 2.1: The Operating Expense Budget

Item	1st Quarter* Actual/Budget	•••	Annual Actual/Budget
Salaries			
Overtime			
Benefit Plans Expense			
Payroll Taxes			
Materials/Suppliers			
Printing, Stationery and Office Supplies			
Maintenance & Repair			
Travel			
Non-Capital Equipment			
Books, Pamphlets, Periodicals			
Conferences and Conventions			
Staff Training			
Telephone (Local)			
Telephone (Long Distance)			
Internal Maintenance			
Building Services			
Shared Administrative			
Duplication Services			
Mail			
Word Processing			
Data Processing			

Figure 2.1: The Operating Expense Budget (cont'd)

Item	1st Quarter* Actual/Budget	•••	Annual Actual/Budget
Consultants			
Production Expenses:			
Writers			
Per Diem Producers			
Per Diem Directors			
External Production			
External Post-Production			
Per Diem Freelancers			
Canned Materials			

*Format would be repeated for 2nd Quarter, 3rd Quarter, etc.

Figure 2.2: The Capital Equipment Budget

Item	Cost	Reason For Purchase	Payback Period
Graphic Arts			
Photographic (Still & Slides)			
Cinemagraphic			
Video Production			
Audio Production			
Multi-image			
Videodisc			
Teleconferencing			

The Operating Budget

For purposes of convenience, the operating budget can be divided into various sections:

(1) Employee Expenses
(2) Direct Expenses
(3) Shared Expenses
(4) Billings

Employee Expenses

Employee expenses can be subdivided as follows: employee salaries; overtime expense; benefits plans expenses and employee payroll taxes. These are relatively self-explanatory expenses.

Direct Expenses

Direct expenses may include a whole host of items, such as materials and supplies; printing, stationery and office supplies; maintenance and repair expenses; travel expenses and external resources.

One might also want to include a budget line here for non-capital equipment expenditures (items that cost less than $500); the demarcation line may differ from organization to organization.

Another group of direct expenses falls into the general area of professional development, such as books, pamphlets and periodicals; conferences and conventions and staff training.

A budget amount for the purchase of outside services, such as temporary office help, might also be included. Some organizations include a budget line for covering errors incurred by the operation. For example, say the slide-making department makes a mistake producing a presentation and the work has to be redone. The cost of that mistake would be placed in the "reclamation of errors" budget line.

Shared Expenses

Shared or "overhead" expenses would include such things as telephone; rent; internal maintenance and repair; building services (such as cleaning); shared administration expenses; use of internal duplication services; use of internal mail services and use of internal word processing services.

Some organizations' media operations might consider paying for their share of the amount of electricity (for electronic equipment, for example) and water used (for the photography department, for instance).

Billings

The last item on the operating budget—Billings—is the bottom line, so to speak. It is this item which will determine how profitable or unprofitable the organizational media operation is. The total annual amount of billings will depend, of course, on a variety of factors. Among these are the rate card developed for the media operation (see also Chapter 6) and the volume of work. Billings, then, refers to the amount of money (sometimes referred to as "funny money") charged to internal clients for media services.

If billings exceed the total of employee salaries, direct expenses and shared expenses, then the organization is running a profitable media department. If the billings more or less equal the total of employee salaries, direct and shared expenses, the operation breaks even. However, if billings run short of the total of all expenses, then the organization is, to a certain degree, carrying the department as excess overhead. In some organizations the excess is divided equally or divided according to a preset formula among all operating divisions.

The long-term survival of the media operation depends, of course, on a variety of factors. But one of the more important factors is the parity between expenses and billings. At the very least, a break-even operation is one that shows that the media department is paying for itself, that there is demand for its services and that internal clients want those services and are willing to pay a price for them.

The Capital Budget

A capital budget represents all the hardware a media department might require, ranging from special filing systems for graphics, to video cameras and computer editing systems.

The equipment to be itemized might include:

- audio recording and mixing equipment
- cinemagraphic equipment
- distribution amplifiers
- film audio recorders
- film projectors
- graphic design tables
- graphic storage
- lighting instruments
- multi-image programmers
- photographic processing
- slide duplication cameras
- slide production stands
- still photography cameras
- 35mm slide projectors
- video cameras
- video editors
- videotape recorders

The capital equipment budget might also take facilities into account. For instance, a photo lab will have special requirements for water and chemical drainage. A video operation might have special requirements for studio ceiling heights, air conditioning and perhaps computer flooring in editing rooms. A graphic arts group may have demands for special lighting. Heating, ventilating and air conditioning requirements (HVAC) may also have to be customized, especially in the electronics part of the center. All of these aspects need to be worked out and developed within the capital equipment budget.

DEVELOPING A BUDGET

The development of an annual budget for an ongoing media department is easier than developing a budget for a new one, primarily because an ongoing operation has a history; year-to-year comparisons can be made, percentage increases in budgets (or decreases, if necessary) can be more easily justified, etc.

Employee Expenditures

So let us talk for the moment about developing a budget for a brand new media department, starting with employee salaries. The amount of employee salaries will depend, of course, on the result of the needs survey discussed in Chapter 1. These kinds of questions will have been answered: How many producer/directors will be needed? How many photographers? How many graphics artists? How many video engineers and/or technicians? To a degree the salary level of each of these production professionals will be determined by various factors: the supply of qualified professionals in the area; the specific region of the country where the organization is located; the general level of salaries for qualified professionals in the country; negotiations between prospective employees and the organization and the general salary structure of the organization.

Expenses for employee benefits plans and payroll taxes will probably be the easiest budget items to determine, primarily because the organization should have already developed formulas for determining these. The formulas will vary depending on the kind of organization and the state and locality in which the organization is situated.

Determining Direct Expenses

A certain amount of guesswork will be made with respect to direct expenses, especially if the organization has no previous media production history. For example, let's take materials and supplies. This budget item might include videotape stock, photographic paper and processing materials, slide mounts and 35mm film, paper, ink, audiotape stock, bulbs and so on. Again, the amount of materials needed in this area will be an outgrowth of the needs analysis performed earlier. One source of information for determining these particular budget items is an outside professional in the business. During the early stages of developing an organizational media production operation, the organization might consider hiring a consultant to help put together the first year's budget. (Consultants are discussed further in Chapter 10.)

On the other hand, professional development expenses should be fairly easy to determine. Relevant trade periodicals and associations are listed in Appendix III. A rundown of the ones that would be appropriate for your organization along with a few phone calls should quickly determine how much you will need for periodicals, conferences and employee professional development.

Determining Shared Expenses

Administrative history can be a guide with respect to shared expenses. Telecommunications management should be able to determine the yearly telephone expenses, depending on how many personnel will be in the department for the year, how many will have one or two lines, how much long distance telephone calling can be expected and so on. Rent should also be a rather simple matter. Presuming the organization has determined a square footage rate for rent, once facility requirements (both office and production) have been established, it should be a simple matter of determining how much will have to be budgeted for rent. Internal maintenance and repair work may require some guesswork but in the overall context of the larger budget a 10% miss here would not be devastating. One can determine a reasonable amount for this budget item by inquiring what internal maintenance and repair work rates are. Cleaning services can be determined likewise. Shared administrative expenses may be dictated by the needs and wants of higher management. Last, internal duplication, mail and word processing services expenses can be determined by finding out what the rates are and estimating the usage of these internal services.

Researching Other Sources

Costs for staff, hardware and direct expenses change from time to time, and differ depending on what part of the country the organization is in. Thus, while experience may be a good starting point for the determination of both the operating and capital equipment budgets, there are various sources one can research in order to get an idea of prices for such items as consultants, production facilities, production houses, free-lancers, hardware vendors and suppliers.

Trade publications are a good source for finding people to talk to regarding prices. They often publish lists of facilities or hardware with the latest list prices. Professional organizations might be useful for determining the range of salaries for certain professional positions. Occasionally, a professional organization will publish a salary survey. The International Television Association (ITVA), for example, publishes an annual Salary Survey. Media production managers at other companies nearby might also be helpful in getting a handle on what prices should be. Hardware companies provide press releases, brochures, catalogs and other information about their products.

THE EVOLVING BUDGET

The dollar size of the budget will depend upon the kind of organization the media production operation is in, the size of the staff, the size of equipment and facilities required and the volume of work. Moreover, the budget should be ready to accommodate ad hoc decisions that might be required during the course of a year that will enable media management to expand (or modify) activities without having to go to higher management to make changes.

A well-prepared budget that is both complete and flexible will give media management the ability to make decisions more independently of higher management. To repeat, a budget is not an instrument of dictatorship; it is, rather, a tool for management to plan, coordinate and control. Higher management will expect the media manager to use his/her budget judiciously. This may mean the manager will have to make decisions during the year that go beyond the original intent of the budget, but still remain within the larger scope of the media department's operations. Therefore, monthly or even quarterly budget reports on how much has been spent and how much has been charged to clients is an ideal time frame within which media management can track progress.

In the final analysis, while a media operation may receive dozens of awards and be considered effective and efficient, the survival of the operation depends largely on economic factors: does it make a profit, does it break even or does it cost the organization money? From the outset, it should be the goal of the media manager to develop a break-even operation. At a break-even point there can be very little argument: the media department is paying for itself, ergo, it has value to its users.

However, organizational management may take the position that, for at least the first year or two, these services can be provided freely to organizational clients until some history can be developed, expenses determined and the volume of potential business takes shape. Once this phase has been passed, there is no reason why an organization should not start charging for media services. While this will not hold for every organization, it seems that the more services media departments charge for, and the higher they charge, the greater the esteem media departments are held in and the greater their volume of work. Perhaps it is the nature of business. Charging for services seems to add a higher value to these services than offering them for free. In the long term, organizations should be charging full-fare for media production services, and the rate of the charge should be whatever the market will bear. (The rate of charges will be discussed in Chapter 6, "Day-to-Day Operations.")

FOOTNOTES

1. Robert E. Seiler, *Elementary Accounting: Theory, Technique, and Applications* (Columbus, OH: Charles E. Merrill Publishing Co., 1969), p.635.

2. A good text on this subject is: Howard S. Noble and C. Rollin Niswonger, *Accounting Principles* (Cincinnati, OH: South-Western Publishing Co., 1961), pp.621-22.

3

Organizing the Media Production Function

The proper organization of a corporate media department can be just as important to its effectiveness as a well-planned budget. The central questions to focus on regarding the organization of media production functions are:

(1) What activities define media production functions?
(2) Which media production functions belong together?
(3) Where in the organization does the media department belong?

DEFINING MEDIA PRODUCTION

The media production operations that define the organization's media functions, are:

- teleconferencing
- videodisc production
- multi-image production
- video production
- audio production
- film production
- computer graphics
- slide production
- still photography
- graphic arts
- overheads
- print

These media reflect the various technologies man has developed for communications: speech, written symbols, printing, still photography, film and slides, broadcast television, multi-image, videodisc and teleconferencing.

Looking at an organization's use of communications media is like looking at a cross section of the Grand Canyon: there are many layers. The older ones have been around for many years. They have not disappeared, and they are still serving a function. In a manner of speaking, they support the newer layers (in this context, the newer technologies).

New technologies are additive; they seem to have the ability to use the characteristics of older technologies. For example, written symbols (such as the alphabet) are representations of speech; the content of printing is written symbols; the content of photography is graphic renderings (painting, for example); the content of film is still photography (film is a series of still frames—photographs—moving at 24 frames a second); the content of electronic communications (such as radio, audiotape, television and videotape) is all the other previously mentioned media. You can put a photograph, slide or film into television (but you cannot put a television program into a slide or photograph). Multi-image by definition implies that it uses all the other media. The current major differences between electronic media and multi-image presentations are of scale, portability and accessibility. Not everyone can communicate via a multi-image presentation; a television presentation is accessible to a lot more people. The videodisc likewise can absorb all the communications characteristics of speech (audio), print, photography, film, television and multi-image. The videodisc also allows the user an effective random access, "interactive" feature not typical of other media.

Interrelationships Among Production Functions

It would be impossible for the various organizational media production functions to operate totally independently. Sooner or later one production function will use or work with another. The frequency of the interrelationship will depend on the nature of the organization, the expertise of those involved in media production, the availability of hardware and, the manner in which media production functions are organized.

For example, a house organ editor may regularly require the services of the graphics arts group for layout and design, as well as the still photography group for originating and developing still photographs. A manager in need of a slide presentation requires the services of the slide production group, as well as the graphic arts department, for preparing the mechanicals to later be presented in slide format. If a manager wants to produce a film, he/she will require the services of a producer/director, as well as the expertise of cinematographers from the photography unit; the slide group may also get involved if photo animation sequences (using slides as the raw material) are required; and the graphic arts group may also be needed if titles or storyboards have to be prepared.

A video production may require the services of all other media production functions, including engineering and technical support; a producer/director; a still photographer for the origination of slides and/or photographs; a cinematographer for the creation of film footage; a graphic artist for the design and execution of titles and storyboards (even set design) and the audio production unit (for the creation of sound tracks involving music-mixed/voice-over and music sound effects tracks). A multi-image production would likewise require the services of all media production groups for obvious reasons.

The more advanced the technology, or the more recent the communications technology required, the more complex the execution of that communication becomes, thus requiring the services of more parts of the media production operation.

Interface of Production and Presentation Modes

Table 3.1 illustrates how the mode of production interfaces with the mode of presentation. As you will note, the more traditional communications media (graphic arts, photography, slides) have a higher degree of interface with the other presentation media, as opposed to the newer media of film, video and multi-image. Graphic arts is the one media production origination group that interacts with all the other presentational media.

THE STRUCTURE OF MEDIA PRODUCTION OPERATIONS

The various media production operations fall into three groups: electronic media, film and graphic arts:

(1) Teleconferencing, videodisc, multi-image, video, computer graphics and audio production can be considered one group, *electronic media*, since electronics is at the heart of these communications presentations.
(2) Cinematography, slides and still photography belong in the film group primarily because *film* in various formats (super 8mm, 16mm, 35mm transparency and still photography) is the technological heart of the activity.
(3) Graphic arts and printing become a third group, primarily because these functions are print oriented.

Table 3.1: Production/Presentation Interactions

	Presentation Mode								
Production Origination	**Telecon-ferencing**	**Video-disc**	**Multi-Image**	**Video**	**Audio**	**Film**	**Slides**	**Photo-graphy**	**Graphic Arts**
Teleconferencing	X								
Videodisc	X	X							
Multi-image	X	X	X	X		X			
Video	X	X	X	X		X			
Film	X	X	X	X		X			
Audio	X	X	X	X	X	X			
Slides	X	X	X	X		X	X		
Photography	X	X	X	X		X	X	X	
Graphic arts	X	X	X	X		X	X	X	X

Several organizational elements remain to be considered: management; the differentiation between software professionals (producer/directors) and hardware professionals (engineers and technicians) and the administrative element.

The question of management is straightforward. Adding this necessary element to the three just discussed results in a basic organizational chart such as the one illustrated in Figure 3.1.

Differentiation Between Software and Hardware Professionals

Within each group is the need for a differentiation between software and hardware professionals. For example, a video producer/director is a software professional, while a video engineer is a hardware professional. Graphic artists, too, can be considered software professionals. However, photographers (whether for film, slides or stills) present a bit of a problem; those involved in originating the production elements of photographic (film) media are both software and hardware professionals. In the case of 35mm slides and still photography, the photographer is often conceptualizer, originator and finisher.

A producer/director is necessary for the development of a film, audiotape, video production, multi-image presentation, videodisc or teleconference. The development of a slide presentation, still photography elements or a graphic presentation may not necessarily require the coordination of a producer/director. Whereas in graphics the graphic artist can be both the designer and executer of the presentational element, the execution of a professional video program cannot be carried out by one person.

Thus, if producer/directors are to be effective in the media production organization, they should be placed where the highest level of media coordination is likely to occur: in electronic media production. Because this group will also require the heaviest investment in hardware and facilities, the hardware professionals required should also be put in this group.

There is also a natural split in photography, between slides (if for no other reason than sheer volume) and still photography. Figure 3.2 depicts the organizational chart as it now looks.

The Administrative Function

Within the media department's structure, one last group is needed for such activities as billing, processing invoices, tracking costs, providing audio visual equipment loan services and conference room setups, meetings services, media distribution and media library services. Thus, an administrative group is formed to round out the centralized media department structure. The completed organizational chart for the media operation is shown in Figure 3.3.

Figure 3.1: Basic Organizational Chart

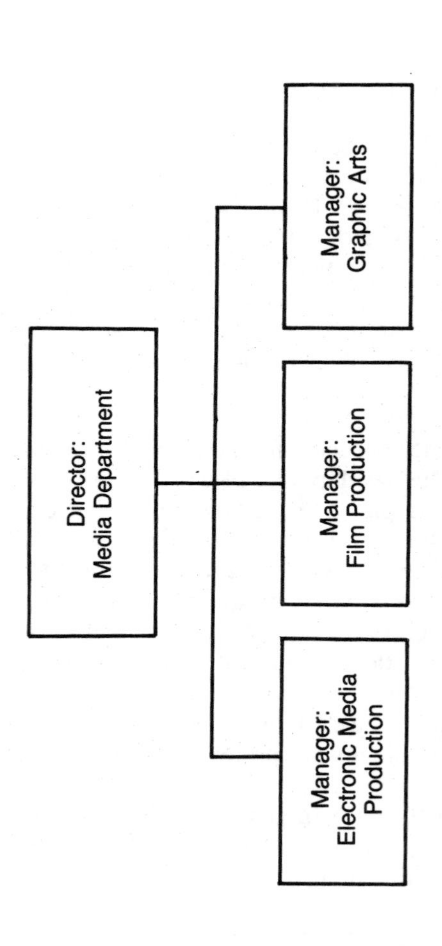

Figure 3.2: The Intermediate Organizational Chart

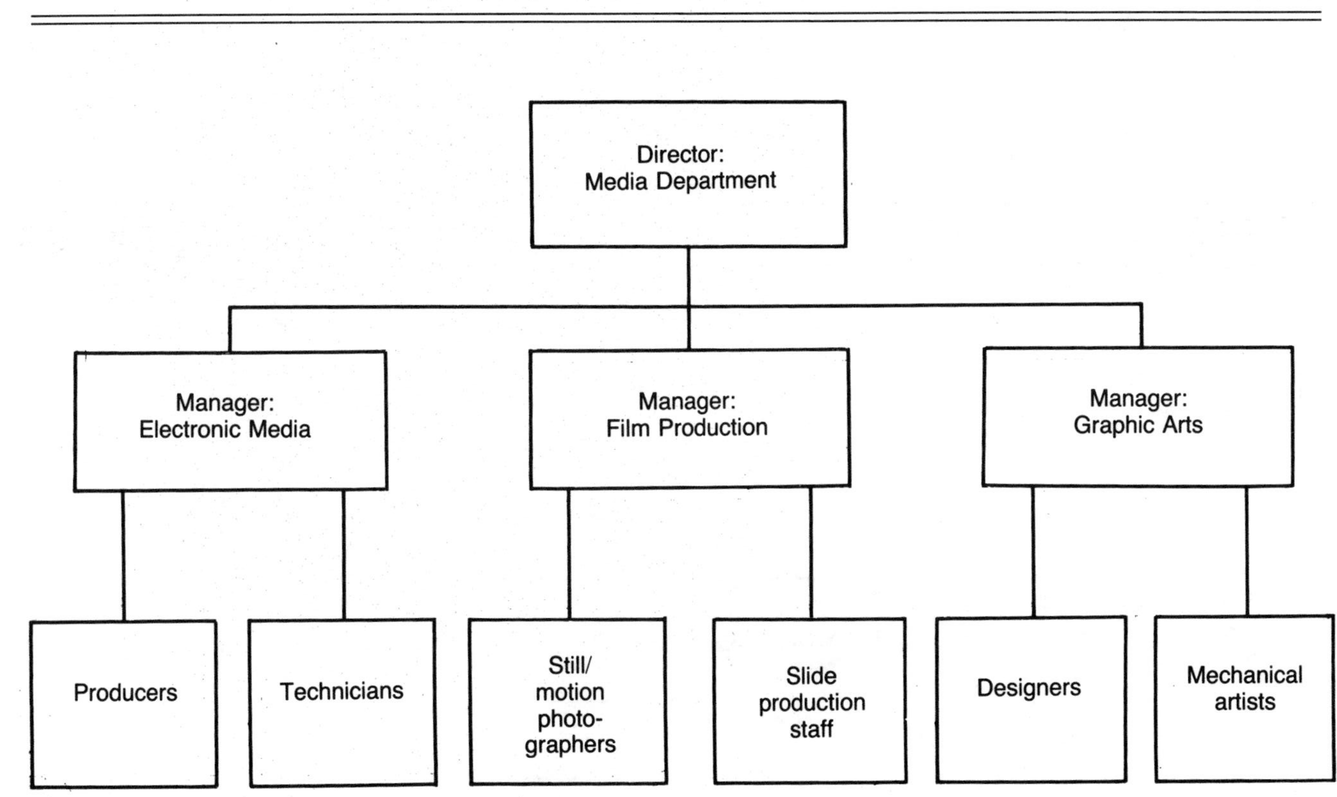

Figure 3.3: Organization of the Media Department

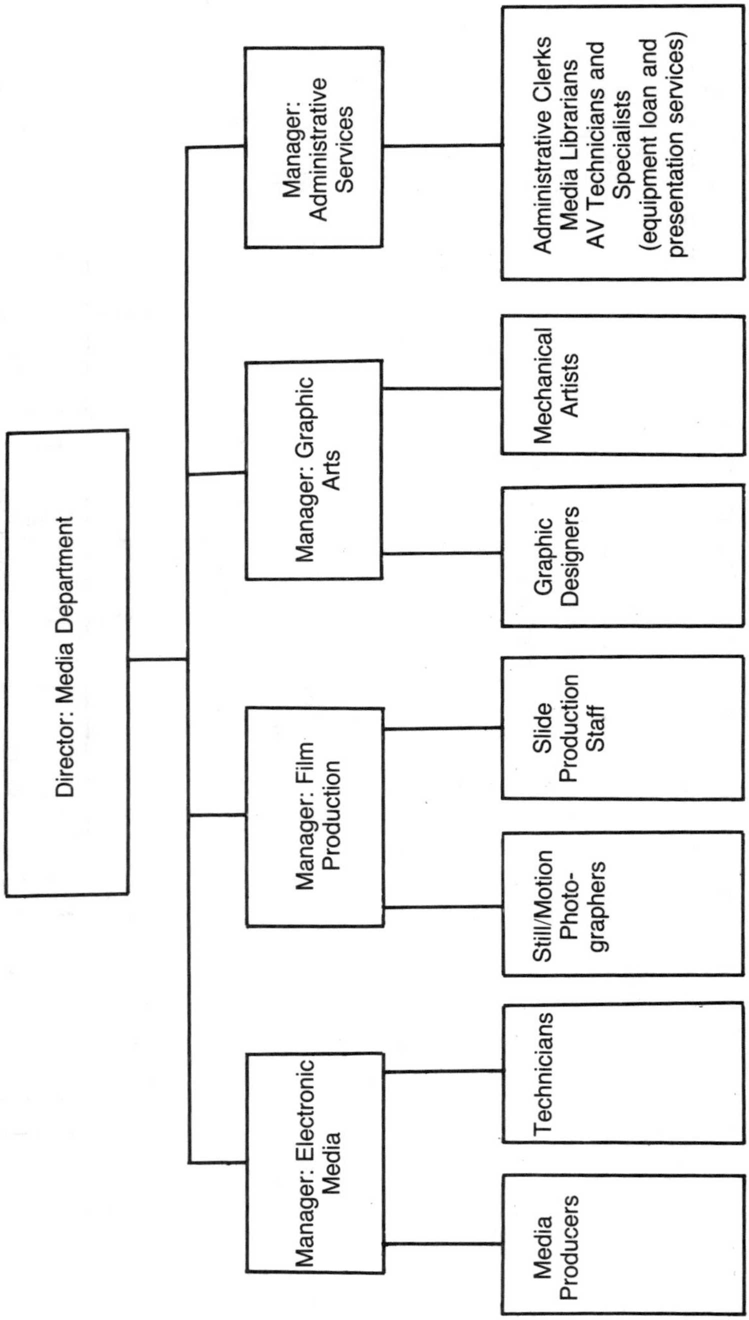

- Director: Media Department
  - Manager: Electronic Media
    - Media Producers
    - Technicians
  - Manager: Film Production
    - Still/Motion Photo-graphers
    - Slide Production Staff
  - Manager: Graphic Arts
    - Graphic Designers
    - Mechanical Artists
  - Manager: Administrative Services
    - Administrative Clerks
    - Media Librarians
    - AV Technicians and Specialists (equipment loan and presentation services)

A centralized approach to the organization of corporate media activities (such as the one given in Figure 3.3) seems to be the best one to take. This is supported by the results of a survey of 50 United States and eastern Canadian companies I conducted several years ago.[1] The survey intended to assess where and how nine "communications functions" were organized. The functions were: (1) video production; (2) printing; (3) graphic arts/photography/AV production; (4) training; (5) press relations; (6) government relations; (7) employee communications (benefits information and house organs); (8) advertising; and (9) community relations. A second purpose was to ascertain if there was an ideal model for the organization of communications functions.

Those responding to the survey strongly indicated that the "ideal" model for the greatest efficiency and effectiveness of communications functions was a well-centralized model. Two findings of interest are:

(1) Printing production was seen by 65% of those surveyed as a function of administration, rather than of communications.
(2) The video and graphics functions were organized into the same department in 60% of the corporations perceiving their model of communication functions as the ideal model, or close to it.

WHERE THE MEDIA DEPARTMENT BELONGS

One of the most important decisions to make concerning the media department is where it will be placed in the overall corporate structure. A media operation should be positioned so that it provides effective and efficient production services to all aspects of the organization, not just a few. For example, if a video group operates under the aegis of a training department, or one aspect of training (e.g., sales), it is likely that video will be perceived only as a training tool. Similarly, if the video operation is situated in an employee communications department, the likelihood is that video is perceived primarily as an employee communications tool. Yet video should not be thought of as only a training tool, or only an employee communications tool; it is both. In addition, it is a tool for management communications and sales promotion and can be used for many other functions such as recruiting, security analysts relations, community relations, etc.

Similar arguments could be made for graphics, still photography and slides. There is, though, a very real question of what is appropriate. People wouldn't get much work done if they had to go to a centralized duplication department that was five floors away and fill out a form in triplicate every time they needed a photocopy.

Each organization has its own requirements, according to its size, geographic location and function. Using video as an example, in many large organizations centralized video production operations may exist at the headquarters level. Smaller video operations may also be located at regional offices, serving specific functions, such as training.

Ideally, it makes sense for a media production operation to be positioned in the organizational structure so that it will be perceived as an organization-wide resource, accessible to everyone and every department in the company. In some organizations this might mean the media department is part of a "communications" content department, such as public relations, corporate communications or marketing. In other organizations, the media department might better serve the company if it is placed in a "non-communications" content department, such as General Services.

Moreover, questions of volume, unit cost and accessibility come into play. Organizations are probably better off having various overhead projectors available for use (and the equipment required for making the overheads) on a decentralized basis because of the nature of the use of the overhead medium. On the other end of the scale, a centralized teleconferencing capability is more cost-effective because of the investment and expertise required for producing a teleconference.

A centralized capability for creating graphics, stills and slides would also make sense for an organization, again for reasons of hardware investment and expertise required.

In the case of computer graphics, a centralized and decentralized operation could exist, primarily because computer graphics technology makes it possible for non-technicians to operate the equipment.

It has become a matter of scale: there are super computers that require very high levels of expertise and there are computers that can fit into the palm of one's hand that can be operated by anyone. This is true of printing, slides, photography and video. The corner print shop can provide quality copies of one's resume at very little cost. But publishing a daily newspaper is quite another matter. Virtually everyone who can afford the cost of a moderately-priced camera can produce slides. But award-winning photos for national magazines require higher-quality equipment and expertise. The home video market has boomed partly because video camera technology is now accessible to the consumer. But videography for a television special demands much more expensive equipment and much higher levels of professionalism.

In the corporate context, virtually anyone can make an overhead and the equipment should be highly accessible. But a teleconference (videodisc, multi-image presentation, videotape or film) needs more than a "home movies" touch. Thus there is a need for a more centralized cadre of professionals and the hardware tools to get the job done.

Figure 3.4: Organizational Communications Problem-Solving Model

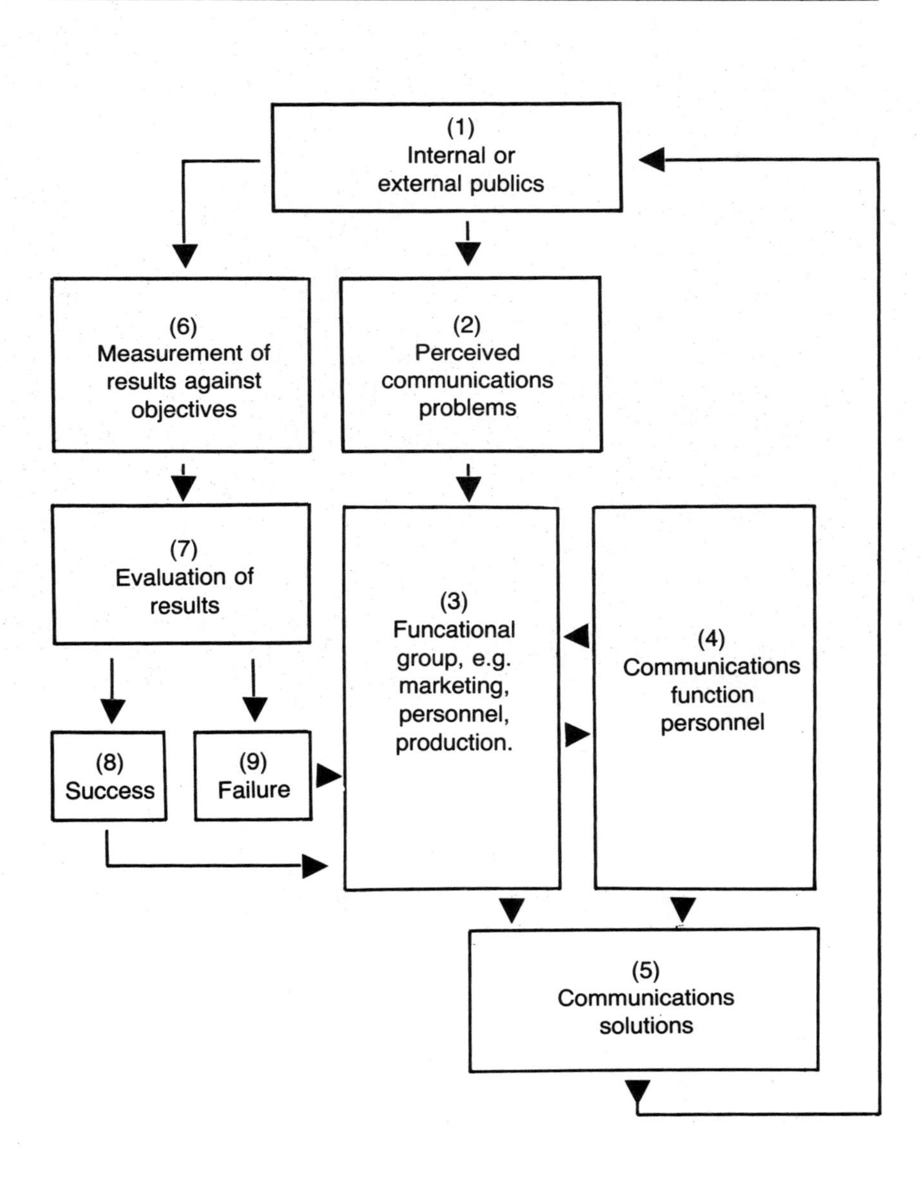

HELPING THE ORGANIZATION SOLVE COMMUNICATIONS PROBLEMS

A media department should help the organization solve its communications problems. Figure 3.4 presents a schematic of the process of communications problem solving in a hypothetical organization. First, the problem is expressed by an internal (employee) or external public (1). The organization, or department, (2) perceives that a communications problem exists. The organization, or department, (3) then seeks resources to solve the problem. It confers with (4) several communications activities in order to get the job done: the graphic arts department, training, video operations, AV services, public relations, and so on. The communications solution (media) (5), whether a teleconference, videodisc, multi-image production, videotape program, audiotape program, film, slide program, photograph, graphic or print piece, is then delivered to the internal or external public (1). These publics react to the solution and the results are measured against communications objectives (6). These results are evaluated (7) as either successful (8) or unsuccessful (9). At this point the results are fed back to the functional group (3), which will have to decide on the need for future action.

The primary role of the media department in this system is to help translate the perceived problem into a media solution that will bring about the desired results.

FOOTNOTE

1. Eugene Marlow, *Communications and the Corporation* (New York: United Business Publications, Inc., 1978).

4

Staffing

One of the hardest decisions a corporate media manager faces is when and how to staff. This chapter offers some guidelines for staffing the media department with respect to what types of positions need to be filled; the qualifications necessary to each; where to find the appropriate personnel and how to keep them.

IDENTIFYING THE POSITIONS TO BE FILLED

The number and type of positions needed in the corporate media department will vary depending on the level of the organization's media production activities. On the simplest level, take an organization in which very little production originates in-house; outside agencies handle most of the work. In this case, the organization should at least designate someone to serve as the media production liaison, or ombudsman. Ideally, this person would be situated in a communications activity. In organizations where several individuals purchase media services from outside agencies, these people should have some experience, and perhaps training, in media production so they can accurately judge the quality of their purchases. Moreover, the organization's central purchasing department must be aware of these activities and monitor the propriety of contractual obligations.

As corporate media production activities increase, so will the need for media professionals. Eventually, the organization may need to consider hiring individuals with expertise in the three major media areas (graphic arts, film and electronic), each of which will be discussed in turn.

Graphic Media

The graphic arts area should be staffed with a professional artist who is adept at both conceptualization and execution. If the volume of graphic arts work is high enough, the organization could consider hiring two individuals: one graphic artist (an art director or graphics designer) and one mechanical artist.

Film Media

Film production could be approached similarly: at the very minimum, a professional photographer should be hired who is skilled in the various film formats starting with still photography (black and white, color and 35mm transparency). Motion photography (cinematography) can be handled by the photographer or free-lancers. The organization might be lucky enough to find an individual with expertise in both still and motion photography. However, while there are certainly common characteristics between the two, there are also many differences, and stretching one individual to handle both kinds of photography could turn out to be penny-wise and pound-foolish.

Electronic Media

The advertisements of hardware manufacturers to the contrary, electronic hardware (such as television cameras and videotape recorders) of even the simplest design cannot be operated or maintained by just anyone. If an organization has enough work to warrant the purchase of video production and post-production equipment, it should staff the electronic media production section with not only one qualified producer/director, but a video engineer or technician as well.

The software (producer/director) and hardware (engineering) functions should be separated in the electronic media section. One reason is that the higher the level of complexity on the technological scale, the higher the degree of specialization required. While electronic media production equipment (cameras, for example) and post-production editing equipment have become more accessible to so-called nonprofessionals, their setup and maintenance remains a problem. For example, there are television cameras on the market today that, with the proper sequence of switch flipping, will go through various setup steps and be ready for use. However, if the camera breaks down in any way, a professional engineer and/or technician needs to be on hand to fix the problem. The same is true of computerized electronic editing systems.

Further Staffing Considerations

There is another consideration with respect to staffing. Many organizations stretch their in-house media staffs by having personnel double up on tasks: graphic artists perform film production tasks, film production personnel are asked to get involved with electronic media production, etc. In other organizations, personnel in communications activities such as employee relations or corporate communications are asked to perform media production functions part-time. Experience has shown that media production activities require attention on a full-time basis, if the final product is to be professional. Moreover, while there are similarities among the three production groups, there are many specific skills requisite to each area; imaging for print, for example, is quite different from presenting a visual image in film or in electronic media. Furthermore, the mechanics for producing a media product for print, film or video are quite different. Finally, it is the very rare individual who can go from being a still photographer in the morning to a video producer/director in the afternoon and still turn out a professional product.

Importance of Specialization

As the media department's workload increases, and as additional personnel are hired, the more specialized each function should become. For example, in a very active graphic arts group, the organization may be required to have several graphic designers and mechanical artists on staff. The graphic arts group may designate specific personnel to handle particular applications, such as promotional literature or communications. Others may be specifically assigned to design graphics for slides.

The film production group may similarly tend to specialize, with some individuals assigned to studio photography and others to field photography. Still others can be designated to handle lab work, such as film development and duplication. The slide production group may follow a similar pattern, with divisions between personnel using slide production origination equipment and those developing and mounting the slides. Depending on the organization, this section might also require one person who handles the client/production interface—perhaps the manager in charge of slide production, or an assistant manager.

As the electronic media production group widens its scope of activity, it too may need more specialized personnel. For instance, in addition to a producer/director and a video engineer, this unit may eventually require video technicians, audio engineers and technicians, production assistants, and so on.

The administrative section could expand to include AV specialists to handle media presentations, librarians to develop and manage a media library and one or more administrative assistants to handle the flow of paperwork associated with invoices, billing and scheduling media production activity.

Thus the number of potential positions in a media production function ranges from one to several dozen, depending on the level and complexity of the media production department's activity.

QUALIFICATIONS OF MEDIA PERSONNEL

The Media Manager

What is a media manager? What role does he/she take on that is substantively different from those of graphics artists, mechanical artists, photographers, producer/directors and the various technicians associated with the media operation? Moreover, what are the characteristics of an effective media manager?

To begin with, media managers share characteristics with their counterparts in other areas. In *The Effective Executive*, Peter Drucker offers five characteristics of an effective executive:

> (1) Effective executives know where their time goes. They work systematically at managing the little of their time that can be brought under control.
>
> (2) Effective executives focus on outward contribution and gear their efforts to results rather than work. They start out with the question "What results are expected of me?" rather than with the work to be done, let alone with its techniques and tools.
>
> (3) Effective executives build on strengths — their own strengths, the strengths of their superiors and subordinates and on the strengths of the situation, that is, on what they can do. They do not build on weakness. They do not start out with things they cannot do.
>
> (4) Effective executives concentrate on the few major areas where superior performance will produce outstanding results. They force themselves to set priorities and start with their priority decisions.
>
> (5) Effective executives, finally, make effective decisions. They know that this is, above all, a matter of system—of the right steps in the right sequence.[1]

To these characteristics we can add several others that distinguish an effective and successful media manager.[2]

Entrepreneurial Spirit

An entrepreneur is someone who organizes and manages an enterprise, especially in business, usually with considerable initiative and risk. Other definitions state that the entrepreneur is the one who puts capital, labor and resources together in an enterprise that did not exist before and makes a business out of it.

The element of initiative and risk is a recurrent theme. Of course, not everyone is expected to go out and, at considerable financial risk, pull capital, labor and resources together to start a business. But there is much to be said for the "spirit" of this kind of activity. Being a manager in an existing organization does not mean that risk and initiative are eliminated. On the contrary, in an age of information, where change is the norm, today's manager must have the ability to take risks and some initiative; it is this "spirit" that separates the merely adequate from the outstanding manager. Moreover, this kind of spirit is not just reserved for the highest echelons of the organization, but is requisite to all levels of the organizational structure. Without it, no new product will surface, no department will flourish and no employee will advance very far.

Marketing Skills

A media manager must have the ability to market and promote his department. The manager must be able to convince others (and keep them convinced) that his area has worth and should be supported. This means developing a network of clients (supporters in the organizational/political sense), garnering recognition for employees and what they do and getting people to come back for more. (See Chapter 6, "Day-to-Day Operations.")

Administrative Skills

All media managers must have administrative skills. What is crucial is to have an understanding of the administrative system within the department and the larger organizational system. Each department in an organization has an administrative system: an organization chart; an expense budget; accounts to be kept; forms, files and figures to be organized; performance reviews; budget reviews; quality control checks; planning procedures. Moreover, no department exists in a vacuum; each relates to the larger organizational network. This means the media manager must be cognizant of organization-wide personnel policies (employee benefits, salary ranges, performance rating systems, etc.) in addition to keeping up with management changes and determining the effect of these changes on the department.

Human Relations Skills

The successful media manager must relate well to people. Machines, administrative systems and formal policies alone do not run organizations: people do. The manager who doesn't understand people at an elementary level is doomed to failure. The successful manager will have a deep understanding of people, knowing, for example, that what people say and what they do can be entirely different, and that the causes of people's actions are not necessarily direct.

Technical Skills

Although there are exceptions, most people enter the management ranks from positions where they had mastered a particular skill: accounting, marketing, engineering, communications, media production, etc. It never hurts for the manager to have some technical skills in the area he or she is managing. While there are those that say a good manager can manage anything, a good manager can also fail without an adequate understanding of the technical skills applied to the particular area. If the manager doesn't have at least a basic understanding of the required technical skills (here, in media production), it must be acquired.

Communication Skills

A successful media manager must be able to communicate formally and informally, in writing and verbally. Well-honed writing skills are a must, whether for preparing a short memo or an annual departmental report. The ability to communicate effectively and efficiently, in writing and verbally, can help overcome deficiencies in other areas.

Flexibility

The competent media manager must be flexible. Change is constant and adaptive abilities are what keep the species viable. If the media manager understands change and has developed mechanisms for dealing with it and the stress that normally accompanies it, then this manager can help others adapt to change.

Production Personnel

Production personnel can be divided into three groups: software personnel, hardware personnel and administrative personnel. Software personnel would include art directors, graphic artists, still photographers, cinematographers, slide designers, producers, directors, writers and media production assistants. Hardware personnel include mechanical artists, photolab technicians, audio engineers, video engineers and audiovisual specialists and technicians. Administrative personnel include administrative assistants, clerks and librarians.

As mentioned previously, there are essential differences between software and hardware personnel. In this context, software personnel usually have direct contact with department clients. Art directors, still photographers and producer/directors have the primary task of helping to translate a client's ostensible communications needs (see Chapter 5, "Solving Communications Problems") into a communications form. Software personnel have the responsibility of shepherding a media project from initiation, to completion, to evaluation.

The role and responsibility of hardware personnel, on the other hand, is to support the activities of software personnel by using and maintaining the tools and equipment necessary to execute the communications project, whether graphic (print), photographic or electronic.

Examining the Candidate's Background

There are four areas to consider in choosing the appropriate individual for a position in the organizational media operation:

- Experience
- Education
- Knowledge and skills
- Interpersonal skills

Experience

Experience affords the greatest test of how appropriate a person is for a media production position, whether it is managerial, software, hardware or administrative. An individual with five or 10 years of experience as a still photographer is obviously more skilled than one who has just graduated with a degree in communications arts. However, there are other factors with regard to experience that must be weighed. What is the quality of the individual's experience? Did the individual's skills grow during the last five to 10 years? How effective or successful was the individual at the previous place of employment?

There may be times when an organization chooses to employ an individual with little experience. First, less experienced personnel will generally require less compensation. Second, older hires are not necessarily more experienced in what the organization *needs* than younger ones. To a degree, new communications technologies and the rapid rate of change weaken the argument for experience as a criterion for future performance. On the other hand, experience also brings with it maturity, understanding and judgment, which can be valuable assets in a candidate.

Education and Training

Education is another criterion for selecting personnel. While it is not necessary that all members of the media staff have an undergraduate degree, it would be preferable, particularly for managerial and software personnel, primarily because these two groups will have contact with organizational executives, clients and subject experts. The credential of a college degree provides at least some educational parity between media staff and the organizational clients. What kind of education is preferable? Management personnel are perhaps the most difficult to define, particularly since there may not be a direct relationship between the college degree earned and potential success as a media manager. Training in the software and hardware aspects of various media, as well as in verbal and written communications skills, is helpful. Formal education in business subjects is also useful, particularly in economics, accounting, marketing, sales, computer science, management information systems, personnel administration and organizational theory.

Software personnel should be trained in the various skills for which they are hired. For example, a graphic or mechanical artist should have some formal education in graphic arts, a photographer should have formal training in photography, etc. Producer/directors, on the other hand, should have formal hands-on training not only in television, film or multi-image, but also in written and verbal skills.

Hardware personnel, such as photo lab technicians, should have highly specific formal training in their areas. Video and audio engineers should have formal education in electronics, whether from a high school, college, university, military or trade institution.

Of course, there will be exceptions where individuals demonstrate knowledge of management, software production and hardware without the benefit of a formal education. In general, an individual's formal educational background should be weighed against his/her level of experience and demonstrated skills.

A person's knowledge, skill and abilities, particularly in software and hardware, should be demonstrable: a graphic or mechanical artist should be able to demonstrate these skills; a photographer should be able to display a portfolio of previous work and discuss which camera, what kind of lighting and what type of developing process were used; a producer/director should be able to present a "show reel" of previously produced video, film or multi-image programs and discuss the various software and technical problems associated with each. Naturally, it will help if the candidate is interviewed by someone who has some knowledge of media production, which is why it is important that an organization have at least one person on board with such knowledge.

Interpersonal Skills

Interpersonal skills are essential for any position on the media staff. Even though managerial and software personnel (and to a degree administrative support personnel) will have more contact with organizational executives, clients and subject experts than will hardware personnel, there is no reason for anyone on the media staff to have poor interpersonal communications skills. To a very large degree, an effective media department depends on the level of its people's interpersonal skills (see also Chapter 8). While there is no substitute for software and technical skill, the ability to understand client problems at all levels in the organization, and the ability to work with others, are qualities that help the media department make an effective and efficient contribution to the organization.

Desirable interpersonal skills and qualities might include the following abilities:

getting along with others
relating to management and creative personnel
dealing with subordinates and superiors
common sense, analytical and logical thinking
strong verbal skills
listening
curiosity and creativity
aggressiveness
patience
attention to detail while seeing the larger system
forming cogent arguments
organization
working well under pressure
enthusiasm and motivation
a positive attitude and willingness to work
being a self-starter
being dynamic, yet mature
a sense of showmanship, where appropriate
a quick study

This emphasis on interpersonal skills and personal qualities may seem overdone. However, the functional objective of the media department is to provide the organization with effective and efficiently produced communications products. Therefore, media staff should reflect a high degree of communication skills when dealing with executives, clients, subject experts or outside resources.

WHERE TO FIND STAFF

Promoting from Within

There are various sources for finding potential media staff. The first and most obvious is the organization's internal staff. Graphic artists could be found from the mechanical artists' ranks; photographers could come out of the photo lab; producer/directors could be promoted from the ranks of technical or production assistants.

External Resources

Another source is other organizations. As a result of the growth of media production departments, increasing numbers of organizations can provide a pool of candidates who are already experienced in media management, software or hardware-related activities.

Executive recruiters and employment agencies can also be a source of potential candidates. On the positive side, headhunters may have more immediate access to appropriate candidates. On the other hand, they will charge a fee for their services, usually a percentage of the annual salary agreed upon between the hiring organization and the candidate.

Professional organizations may also provide leads (see Appendix II for a list of professional organizations). Some professional organizations maintain a job-matching service for members. Sometimes, those with newsletters regularly or periodically publish lists of available members for employment. Organizations looking for potential candidates can often publicize themselves in these newsletters.

Advertising in professional or trade journals is another route. Most communications and media production professionals read at least one professional or trade journal. The candidate the organization is looking for might be found by advertising in these publications (a listing of professional and trade journals is found in Appendix III).

Educational institutions, such as high schools, colleges or universities with specific communications programs (either software or technical) or trade schools can also be an invaluable source for potential staff. Very often these institutions provide an employment service for students to match potential employers with graduating students and alumni. In addition, some institutions seek to have their students employed as interns.

Organizations can also find candidates by advertising in widely-read publications, such as the *Wall Street Journal, The New York Times* or regional and local publications.

MANAGING THE STAFF

In addition to staffing the media department with qualified people, the successful media manager must also consider some of the aspects of managing creative media production personnel. First, it helps to think of creativity in terms of three main elements: an ability to make order out of chaos; an ability to put two unlike things together which result in a third thing; and knowing how to take this new thing and communicate it in its simplest terms in a way that it is both informative and entertaining. The major difference between the designations of creative and noncreative personnel is that the former involves more highly developed skills based on inherent abilities. It is really a matter of degree, not kind.

The first step toward practical, yet creative, management of the media staff is to make order out of chaos. Job descriptions are essential. They give the manager and staff a starting point for understanding performance. A second step is the creation of performance standards. While a job description is a word picture of what the job is, a performance standard is a word description of what the individual will accomplish over a certain time period. Both documents serve as a dynamic, evolving foundation for the work to be performed.

A third and important aspect of managing the creative staff is to provide adequate compensation. Creativity might provide psychological compensation for a while, but sooner or later the good staff will need monetary compensation as well. Creative personnel also enjoy the feeling that there is a chance for upward mobility. If a career ladder doesn't exist, create one. That young AV technician pushing around slide projectors may one day become a senior production supervisor.

Apart from these formalities, the most important service a media manager can provide for the staff is an environment for the development of creativity. This means that the manager must have a feel for the creative process. Moreover, the manager must know when to challenge fiercely and when to lay off. Each creative person is different and even creative geniuses have dry spells.

Exchange of information is a natural part of the creative environment. Staff meetings are an ideal way for creativity to be stimulated, honed and channeled. Providing the opportunity for staff to attend professional conferences and seminars is another way. All these kinds of information-gathering activities help the creative person to expand his/her horizons, to grow, develop and make an even larger contribution to the unit. The manager needs to encourage the creative staff to continue to seek new information, as well as to develop interesting and effective new ways to communicate it.

FOOTNOTES

1. Peter F. Drucker, *The Effective Executive* (New York: Harper & Row, 1967), pp. 23-24.
2. Some portions of this chapter are adapted, with permission, from Eugene Marlow, "To Manage or Not to Manage," *Videography*, February 1980, Copyright United Business Publications, Inc.

5

Solving Communications Problems

For a corporate media department to be successful, it must have an operating approach that makes it a contributor to the overall and long-term development of the organization. To be effective, the media staff should adopt a systems perspective of the organization. It is imperative that, regardless of the function or position, media personnel understand that they are not working in a vacuum, but are performing a service within the context of the larger corporate training, employee communications and external communications system.

Successful communication is not easy. There are at least nine reasons why communications, no matter what the medium, fail:[1]

(1) Failure to define the communications problem.
(2) Failure to define the objectives of the communication.
(3) Failure to measure the results of the communication against the objectives.
(4) Failure to profile the intended audience.
(5) Failure to take into account where the intended audience is and in what environment it will receive the communication.
(6) Failure to determine whether the communication will prove cost-effective.

(7) Failure to ascertain which medium (or media) will be the most effective.
(8) Failure to define how the program will be produced (pre-production planning, production, post-production, duplication, and distribution).
(9) And, most important, failure to evaluate whether the communication succeeded or failed, and why.

Turning these negative statements into positive questions, we can develop an approach to communications problem solving by asking:

(1) Is there a communications problem?
(2) What are the objectives of the communication?
(3) How will the results of the communication be measured against the objectives?
(4) What is the profile of the intended audience?
(5) Where is the intended audience and in what environment will it receive the communication?
(6) Will a formal communication prove cost-effective?
(7) What medium (or media) will be the most effective?
(8) How will the program be produced?
(9) Did the communication succeed or fail, and why?

IS THERE A COMMUNICATIONS PROBLEM?

In most instances, clients come to the media production department with problems that are truly communications-related. In certain cases, however, the communications needs may be more apparent than real; i.e., media may not be the answer to the problem. Blindly producing communications programs for apparent communications needs only serves to perpetuate the myth that communications in any form can and will solve any problem, and that the mere act of producing and communicating a program will automatically do the job.

Therefore, it is important to know whether the client has indeed come to the right place. In some cases, the answer will be easily determined, either because the communications problem is simple or because the client has adequately determined and defined the problem and objectives. However, if a client is unable to articulate the problem or define objectives, he/she may be seeking a solution to what is not a communications problem. For example, a West Coast company was experiencing a decrease in morale on the part of the sales force, and sales were declining markedly. The company decided that the sales force needed more training and/or some sort of motivation and called in a consultant to prepare a program. After a period of research, the consultant discovered that the problem was really administrative: the salesmen were not receiving their commission checks within a reasonable period of time following the close of a sale. The consultant recommended ways to streamline the system for issuing commission checks. Once the check delivery system was speeded up, morale and sales increased.

DEFINING THE OBJECTIVES OF THE COMMUNICATION

Objectives should be spelled out clearly in specific terms for all to understand. To be effective, the objective must be defined in such concrete language and terms that it will provide a criterion for the measure of success (or failure) of the communication. In other words, objectives should serve as a yardstick: how much money did the program help generate, how much was efficiency increased, how much was productivity enhanced, how much money was saved, etc.

Robert F. Mager suggests an excellent method for defining training and instructional objectives.[2] (While his programmed-learning text is designed primarily for teachers, it can also be used effectively by media professionals.) Mager provides the following structure:

(1) A statement of instructional objectives is a collection of words describing one of your intents.
(2) An objective will communicate your intent to the degree you have described what the learner will be doing when demonstrating achievement and how you will know when he is doing it.
(3) To describe terminal behavior (what the learner will be doing):
   a. identify and name the overall behavior act.
   b. define the important conditions under which the behavior is to occur.
   c. define the criterion of acceptable performance.
(4) Write a separate statement for each objective: the more statements you have, the better the chance you have of making clear your intent.

MEASURING RESULTS

Measuring the success or failure of communication programs (and hence, the clarity and reliability of the statement of objectives) is easier for some programs than for others. Communications content can be defined on two levels: (1) content that makes knowledge more productive—communications that motivate individuals to learn new skills, the results of which can be translated directly into more efficient corporate operations, sales and earnings; and (2) content that creates understanding—communications programs that bring vital information to employees so they have a better understanding of the context in which they are working. The results of communications in the former group are easier to measure and analyze than those in the latter group. However, the effectiveness of both kinds of communications programs must be measured in some form.

To develop yardsticks to measure change against objectives, it is necessary to first assess the current situation. Assume, for example, that a training program is expected to increase production in a manufacturing plant by 10%. One must determine the current level of productivity. In other words, to effectively measure any changes resulting from the new training program, the post-testing results must have a base, or control, from which to draw comparisons. Once this base has been established, one can determine whether the new communications training program has succeeded.

Several kinds of changes can be measured, such as a change in quantity, or a change in attitude. A change in quantity is simply a measure of numbers. For example, analysis may show that *x* number of people produce 20 widgets per hour. After dissemination of the communications program (say an interactive videodisc or videotape program), evaluation may show that productivity is now 30 widgets per hour, an increase of 50%.

When evaluating for a change in attitude, one can use a variety of scales, including a Likert Scale (a five-point scale in which the interval between each point on the scale is equal) or a semantic-differential scale (usually a seven-point scale).[3] The Likert Scale provides responses ranging from "strongly agree" to "strongly disagree"; a semantic-differential scale provides a respondent with bipolar responses (such as "original" as opposed to "conventional"). For example, prior to a communications program, a salesman may be negatively disposed to an impending sales territory reorganization as measured on a Likert scale. Following the dissemination of a communications program (e.g., videotape and/or print) outlining the advantages of the reorganization, a Likert scale would be a useful instrument for measuring any change in attitude.

Change can most easily be gauged, comparatively speaking, for skills training programs, primarily because such programs involve specific steps in a process to be learned. For example, "After the training sessions, the mechanics will be able to turn five widgets in 10 seconds, with a 95% degree of accuracy." Motivation programs are the hardest to measure because the degree of change in motivation may or may not be immediately obvious; the audience may need time to digest and respond to the motivational aspect of the communication.

PROFILING THE INTENDED AUDIENCE

Knowing the market is an inherent part of the marketing process. Rather than products and services, however, media producers market and sell information that (1) makes knowledge more productive, and (2) creates understanding. No organization should attempt to produce communications without first knowing some of the basic facts about the intended audience, such as:

- Male, female or both?
- Age groupings?
- Management, professional, staff, clerical, external?
- Level of work experience?
- How many?

Another aspect to the audience profile question has to do with the nature of the audience's perceptions. One simple technique for determining the audience's perceptions is to seek out a portion of the prospective audience and listen! Almost any worthwhile handbook on business communications recommends this procedure. This technique shows that what the information source (e.g., management) believes are the communications needs and what the receiver (e.g., customers, employees) believes these needs to be, are sometimes two very different things. Questioning part of the prospective audience can help obviate the trap of communicating information in a manner inappropriate to the audience's needs.

One net result of this process is that the audience will tell the communicator how the so-called communication should be prepared. The whole process should result in a more effective communication; i.e., the chances of achieving the defined objectives will be enhanced. Moreover, finding out the facts about the audience (the demographic perceptions) will have a direct effect on the style and language of the communication.

LOCATION OF THE AUDIENCE

The location of the intended audience could have a direct influence on the choice of media. To determine this, certain questions must be considered:

Will an audience congregate in a central location or several locations to receive the program?
Does the presentation have to be portable?
Are there competent individuals available to present the program?
Must the program be available to anyone at any time?
Is there a need to distribute the program in various media?
Will large groups, small groups or both receive the communication?
What is the life of the communication in relationship to the audience?
Will it be received once; several times in a short period of time; in several locations at the same time; or will the communication

be used for several years; or go to several different types of audiences?

Is the audience in the United States or in various overseas locations?

Questions such as these must be answered before choosing the production technique and distribution medium. It may turn out that the production medium and distribution will have to be different. Assume, for example, that the communications program involves sales promotion and that the salesperson must bring the program into a prospective client's office. The media choices will have to match the varying environments. The salesperson would have some difficulty lugging a 3/4-inch videocassette unit and monitor from office to office, whereas a portable super 8mm unit or VHS would do the job admirably.

Knowing where the audience is will affect how many units of distribution hardware will be needed. For example, it may be that a sales force is scattered in a relatively small geographic area and it is important that a subject expert be present following the formal communication. In this case, it could be decided that the formal communication can be hand-carried by the subject expert from location to location over an extended period of time. On the other hand, if analysis shows that formal communications programs will be distributed to the sales force over a period of time, then a more permanent distribution network may be needed e.g., a videocassette or videodisc network.

COST-EFFECTIVENESS OF FORMAL COMMUNICATION

Cost-effectiveness refers to how desirable it is to spend *x* amount of dollars for *y* amount of results. Of extreme importance in the communications problem-solving process, cost-effectiveness is the factor that management and individual clients look at when determining the desirability of producing a formal communications program, or when deciding whether to initiate a broad communications function (e.g., a videocassette or teleconferencing network). The dollar figure will also help determine whether one communications medium should be chosen over another.

In this discussion, the term "formal communications" refers to information conveyed in a structured and succinct manner and does not apply to interpersonal communications in which no formal media presentations are involved. Greeting the secretary in the morning is an informal communication; a 10-minute film on the company's dental plan is a formal communication.

An objective way of approaching the cost-benefit/medium choice is to determine what will be gained by creating a formal communication. There are various angles from which cost-effectiveness can be viewed:

(1) What is the cost per person of one medium vs. another?
(2) What will it cost if content is not formalized?
(3) Will the formal communication help increase revenue?
(4) Will the formal communication help reduce expenses?

Cost-Per-Person Comparisons

One can determine how many people will actually receive the communication over a period of time. If, for instance, a $5000 video program dealing with training skills will be viewed by 10,000 employees over a period of one year, the cost per viewer is 50 cents, plus employee viewing time. However, a cost-per-person analysis is, at best, only a tool to weigh the financial advantages of distribution by one medium as opposed to another. It may be cheaper to distribute the information via print (e.g., a booklet), and it may turn out that, everything else being equal, the ''cost-per'' via print is less than 50 cents. Of course, distribution costs are not the only consideration; if they were, then the cheapest route to follow would be word of mouth (extemporaneous communication) rather than any formal communication. Word of mouth, of course, allows for no control or uniform delivery of information, and could take forever. However, an analysis of distribution costs on a ''cost-per'' basis can be a useful starting point.

Reducing Individual Time Investment

Cost-effectiveness can be viewed from other angles. For example, can a communications program reduce the amount of time needed to learn skills in an existing training situation? Assuming that conventional training methods (stand-up lecture, on-the-job training, programmed instruction booklets, operating manuals) require 10 hours of a student's time, will a new communications program reduce the amount of time invested (with the same degree of results or better) than an existing program? If the answer is yes, then cost-effectiveness is at work. Again, the cost of the new communications program must be compared to the cost-effective advantages gained by using it.

Increasing Revenues

Another aspect to the cost-effectiveness of communications programs is that of profit. Many companies have discovered that communication media can be used effectively for external sales promotion—for instance, when a company wishes to make a sale to the president, chief operating officer, purchasing officer or high-level vice president of another company. Many companies are using portable super 8mm film projectors (with rear and front screen capability) or portable VHS players to make their pitch. Substantial profit can result from an investment in such programs. For example, a sales promotion program costing $30,000 can generate accounts of well over that amount, especially if the market involves the sale of a single product with a minimum cost of $30,000.

Making Better Use of the Time Available

Information derived from a subject expert that runs two hours in extemporaneous form can be condensed into a 20-minute videotape. But does this mean that the company saves an hour and 40 minutes of everybody's time? It may. In addition, the two hours can be put to better use: for example, the 20-minute videotape can be followed by an hour and 40-minute discussion session with the subject expert. The value of a subject expert is not necessarily his/her ability to present valuable information time and again, but his/her ability to answer specific questions. Therefore, when an audience has the opportunity to ask questions and develop a dialogue with a subject expert, the chances are higher that the audience will have a better grasp of the material presented than if no repartee develops. This illustrates that while creating a formal communication may indeed save time and money, its cost-effectiveness does not stop there.

CHOOSING THE MOST COMMUNICATIONS-EFFECTIVE MEDIUM

In addition to frequency, audience size and location, the inherent content of the communication must be analyzed in order to choose the most effective medium:

(1) Does the communication intend to teach motor skills, or attitude/behavior changes?
(2) Is the person delivering the content as important as the content itself?
(3) Is the content of the communication primarily factual in nature?
(4) Is the inherent content of the communication conceptual in nature?

Show Me

Studies have shown that motion is best taught in a media presentation, especially if the information is not easily conveyed verbally or the students are unfamiliar with the task, e.g., operating a machine. This suggests that a film, videotape or videodisc presentation would be more desirable for teaching motor skills than a series of static slides.

These studies also suggest that where imitation is required—for example, learning an attitude via real life demonstration or role models—film, video or videodisc are ideal media for communications effectiveness.

VIPs

Sometimes the person delivering the message is as important as the information itself. For example, executives are inherently important organizational figureheads. What they say is important because *they* are saying it. In this instance, a still slide or photograph together with an audiocassette recording will have less impact than if the audience sees the executives themselves. While a video or film communication is second best to having a live presentation, in-person appearances are not always possible or cost-effective, especially if the organization is geographically dispersed. But video or film remain far superior to a print or other static piece.

Video remains one of the most effective ways of motivating personnel. However, since video is once-removed from personal contact, the type of material presented must be chosen carefully. Too often management falls into the trap of presenting on video the kind of motivational pep talk that may work well person to person, but which seldom transfers to the video medium.

Just the Facts

If the communications problem involves teaching factual information, a person may not need to be part of the communication and motion may not have to be part of the presentation. The learner may be required to digest information in a predetermined sequential manner. That is, first perform step A, then step B, then step C, and so on. Factual information may involve mere numbers, such as those in a stockholders' annual report.

Print in its many varieties, and other static communications media (slides and overheads), can communicate this type of information quite well. Research into the relative communications effectiveness of various media has shown that there is relatively little difference among them in cognitive learning of factual information.

Concepts

Conceptual information is a different matter. One of the earliest and most important uses of video, for example, was to explain a process, particularly a process involving a concept normally difficult to grasp. Films and videotapes containing graphics and/or simple animation can explicate complex ideas faster and more easily than the printed word.

PRODUCING THE PROGRAM

The production process for any medium involves four stages:

(1) Pre-production planning
(2) Production
(3) Post-production
(4) Duplication and distribution

The pre-production stage is where all planning takes place. With respect to teleconferencing, this might mean planning the meeting agenda. For videodisc, videotape, film and multi-image presentations, this step involves developing content outlines, determining where shooting will take place, developing production budgets, selecting necessary per-diem personnel, making travel arrangements, casting, etc.

Production involves creating the actual elements of the program: drawing mechanicals, in the case of graphic arts; shooting, in the case of film and electronic media. During production, raw material is created for later refinement.

Post-production is another term for organizing the raw material created during the production phase. Post-production involves creating layouts and mechanicals in the case of graphic arts and editing in the case of photographic/cinemagraphic and electronic media.

The last stage is duplication and distribution, making copies of approved programs (whether in print, slide, photographic, film, videotape or videodisc formats) and getting those programs to the appropriate audience.

EVALUATION

Once the communications program has been produced and distributed, it must be determined whether all the analysis, time, money and labor spent on producing the program achieved the desired results. Thus it is important to specify yardsticks, such as stated objectives, early in the process.

Communications should be evaluated at various stages of the program's development. Such evaluations may suggest beneficial changes of length, style of presentation, medium of communication, use and distribution of the finished product.

The simplest, yet most informal feedback mechanism would be to ask the client whether he or she felt that the communication succeeded or failed. On a more scientific level, formal questionnaires could accompany the communication program or be distributed to the audience some time after the communication has been received. The questionnaire might involve the use of Likert semantic-differential scales, open-ended questions or yes/no type responses. Questionnaires might also ask how many people at a particular location received the communication, if there were any technical problems and related questions. Questionnaires are further discussed in Chapter 6.

If success has been achieved, go on to the next project! But if the desired results were not achieved, one should reevaluate the whole process from the beginning. Was the communications problem properly defined? Were there weaknesses in any of the decisions made during the development of the communications product?

FOOTNOTES

1. This material is adapted, with permission, from Eugene Marlow, *Communications and the Corporation*, Copyright 1978, United Business Publications, Inc.
2. Mager, Robert F., *Preparing Instructional Objectives* (Belmont, CA: Pitman Learning, Inc., 1975).
3. The reader is advised to refer to a basic research text, e.g., *Conducting Educational Research*, 2nd ed., by Bruce W. Tuckman, Harcourt Brace Jovanovich, Inc., 1978, to learn about these and other measurement methodologies.

6

Day-to-Day Operations

COORDINATION AND SCHEDULING

Regardless of the size and complexity of the media production function, coordination and scheduling are requisite to effectiveness and efficiency. To avoid missed deadlines, scheduling conflicts and just plain chaos, each major unit of the department (electronic media, film, graphic arts and administration) must have mechanisms for scheduling and coordination. One of the most useful of these is a scheduling board.

Scheduling Board

A scheduling board can be set up using horizontal and vertical axes. The horizontal axis can provide time segments: daily, weekly, monthly or a combination. Depending on the volume of work to be handled, the time frame might consist of daily slots for first three months and weekly slots for the succeeding three months. The vertical axis can list the names of the personnel involved in each project and the name of the project. Individuals in the unit could be given color-coded magnetic dots.

Colors can also be used to distinguish among the various phases in a project's development (i.e., pre-production, production, post-production, duplication and distribution). Different magnetic tabs could be used to indicate if personnel are working on site or on location.

This board thus allows the unit manager to immediately see potential scheduling conflicts (such as when the board indicates one person is scheduled to be in two places at the same time), when too much work is scheduled for one week and very little for the next, if one individual is overloaded and others have a light work schedule, and so on.

The board should be placed in an area where all personnel can view it. Properly used, the scheduling board can reassure management that work is proceeding in an orderly and coordinated fashion. The board is also useful as a marketing tool for new clients who, when shown the board, realize that the unit is busy and productive.

Staff Meetings

Regular staff meetings, even if the group is small, are important for coordination. They provide opportunities for the manager to impart information and for the staff to give the manager feedback on successes and failures, to raise questions and to provide recommendations. The staff meeting presents an opportunity for unit personnel to feel they have a stake in the unit's success. Meetings also provide a forum for staffers to discuss current and future projects and the problems associated with them. The meetings can also serve as a time for personnel to review finished projects and receive feedback from the group on their perception of the project. These feedback sessions can be very valuable, especially if the manager wants to create an atmosphere of free exchange where personnel feel they can give and receive feedback without detriment to their egos or careers. This is much easier said than done. But the risk is worth taking, particularly if the manager wants to maintain a dynamic and open atmosphere where ideas are freely exchanged. Ultimately, everyone benefits.

Staff meetings can also be used for periodic broad looks at the progress of the department, perhaps every six months. These meetings might start with a simple history of the operation, a "where did we come from?" scenario: When did the department begin? With what kinds of applications? What was the original staffing and budget? A good analysis can help dispel fuzzy perceptions about the department's progress.

Next, media management might want to take stock of the client list. Who are the clients? Is there one client, several, a dozen? Who are its oldest clients? What about the most recent clients? What programs have been produced? Has there been a shift in emphasis? This analysis might uncover some trends and growth patterns.

Additional analysis might be made of the staff. Are they competent? Partially trained? Ready for new challenges? And equipment—is it up-to-date or obsolete? What is the level of production costs—not just out-of-pocket expenses, but also the time spent on each project (on average) multiplied by hourly rates? How do production costs compare to post-production and duplication costs? Is there an imbalance? What are the average costs? How do these statistics compare to the average length per program? Is there a correlation?

COMPUTERS

A media production operation that is not using computers for running its day-to-day operations or for some aspect of media production is a media operation about to go out of business.

Computers have become a common fixture of the media production workplace. Today, computers allow graphics personnel to produce printed materials (à la desktop) that look exactly like traditional typeset. Computers can create graphics for use in print or slide form, can touch up still photographs and can create graphics for video projection or for use in finished video programs.

Personal computers with business-oriented software can be used to develop budget estimates and content and style proposals, to develop (and revise, and revise and revise) scripts, provide production budget actualities, and so on.

Computers allow the user to manipulate information in ways that are not possible in a more linear, print-oriented administrative environment. The next chapter describes these ways in greater detail.

FORMS, FORMS, FORMS

Paperwork is universal to every organizational media production function. Basically, forms fall into four categories: (1) pre-production, (2) production, (3) post-production, and (4) general administrative. Effectively used, forms can serve a multitude of purposes and provide several benefits, such as the following:

- Serving as a guide to production personnel in the development of a project.
- Establishing a control system from which responsibilities can be delegated.
- Incorporating changes in a project.
- Providing reliable data for effective decision-making.
- Encouraging planning and control of production activities.
- Recording the history of each project and facilitating subsequent estimates of costs and schedules.

Pre-Production Forms

Pre-production forms allow virtually anyone in the department to help a client determine what is required. Pre-production forms (also known as "requests for media production") should contain information that is indicated in Figure 6.1.

Figure 6.1: Sample Request for Media Production Form

Date	Client	Job #
Subject	Project Title	
Project Purpose (e.g., training, employee communications)		
Medium (e.g., videotape, film, audiotape)		
Program Objectives		
Program Content		
Target Audience		
Audience Size		
Audience Location(s)		
Distribution Format		
Due Date	Estimated Budget	
Approval Signatures	Production Schedule	

Pre-production forms should also contain room for a complete production budget analysis. Table 6.1 lists the budget details required to completely cover all the items that could be involved in a multi-image, videodisc, video or film program.

Table 6.1: Items to Include in a Production Budget

Crew	Set Expenses
Producer	Site rental
Director	Carpenters
Technical director	Grips
Lighting director	Electricians
Camera operators	Props
Videotape engineer	Hardware/paint/lumber
Audio engineer	Special effects
Production assistant	
Electricians	
Grips	**Location Expenses**
Makeup	Location fees
Wardrobe	Scouting camera
Scriptwriter	Guard/police
Set design	
Casting	

Table 6.1: Items to Include in a Production Budget (cont'd.)

Travel Expenses
Auto rentals
Air fares
Hotels
Meals
Petty cash

Rental Expenses
Camera
Sound
Grip
Lighting
Generator
Dolly
Videotape recorders

Miscellaneous Expenses
Administrative
Phone/Cables
Messengers
Trucking
Petty cash

Distribution
Videodisc
16mm film
Super 8mm film
3/4" videocassettes
1/2" reel-to-reel
1/2" Betamax or VHS
Slide dups
Audio dups
Shipping

Graphics Expenses
Artwork
Supers
Slides
Stills

Stock Expenses
16mm film/develop
Videotape stock (1/2", 3/4", 1", 2")
Audiotape stock

Talent Expenses
On-camera principals
On-camera extras
Voice-overs
Models

Post-Production Expenses
Film (editors, opticals, animation, special effects)
Videotape (bump-up protections, time coding, shooting artwork, off-line editing, on-line editing, film-to-tape transfers, tape-to-film tranfers)
Stock footage
Music
Sound effects
Narration/overdo studio
Mixing studio
Transcriptions

Budgetary forms for graphic arts and film production units might include items such as:

Initial design and layout
Design revisions
Typography
Illustration
Photography (black and white, color)
Photostats
Word copy
Tabular
Pictorial
Original photography (location, studio, copy, portrait)
Photography size (4x5, 8x10)
Film color negative, color transparency, black and white
Number of originals (flat art, double burns, kodaliths)

Line graph
Bar chart
Pie chart
Compound chart

Processing costs
Printing costs
Proofsheet costs
Duplication costs

The front part of the budget form should consolidate the aggregate costs of the four basic elements of the budget: pre-production, production, post-production, and duplication and distribution. In this way, the client has a clear idea of the basic steps in the production process and how much each will cost. Moreover, each item on the budget form should have two columns: estimated costs and actual costs. As each phase of the production process is completed, the producer/director and the client will be able to track the efficiency of the production process *during* the course of production, rather than waiting until the program is "in the can," so to speak.

Other pre-production forms include storyboards, location survey checklists, lighting plots, set designs, scheduling forms, engineering checklists and production personnel checklists. A storyboard is represented in Figure 6.2. An example of a scheduling form is shown in Figure 6.3.

Production Forms

Production forms consist of documentation taken during the actual making of the project. These include camera shot lists, scripts and release forms.

A script page contains the following information: writer, title of the program, video information and audio information (including on-camera audio, music, special effects and voice-over). Release forms (see Figure 6.4) protect an organization from liability when subjects are used in a photograph, slide presentation, film, videotape or multi-image presentation. The legal department can develop a release form. Chapter 11 discusses releases in greater detail.

Post-Production Forms

Post-production forms can include such documentation as editing worksheets and shot logs. Also in this category are customized labels for videodiscs, videocassettes, audiotapes, films and slide presentations, as well as videotape duplication requests, library request forms, mastertape labels, duplication labels and feedback forms.

Feedback forms help determine if a media project was a success or a failure, or if it achieved a measure of accomplishment somewhere be-

Figure 6.2: Sample Storyboard Layout

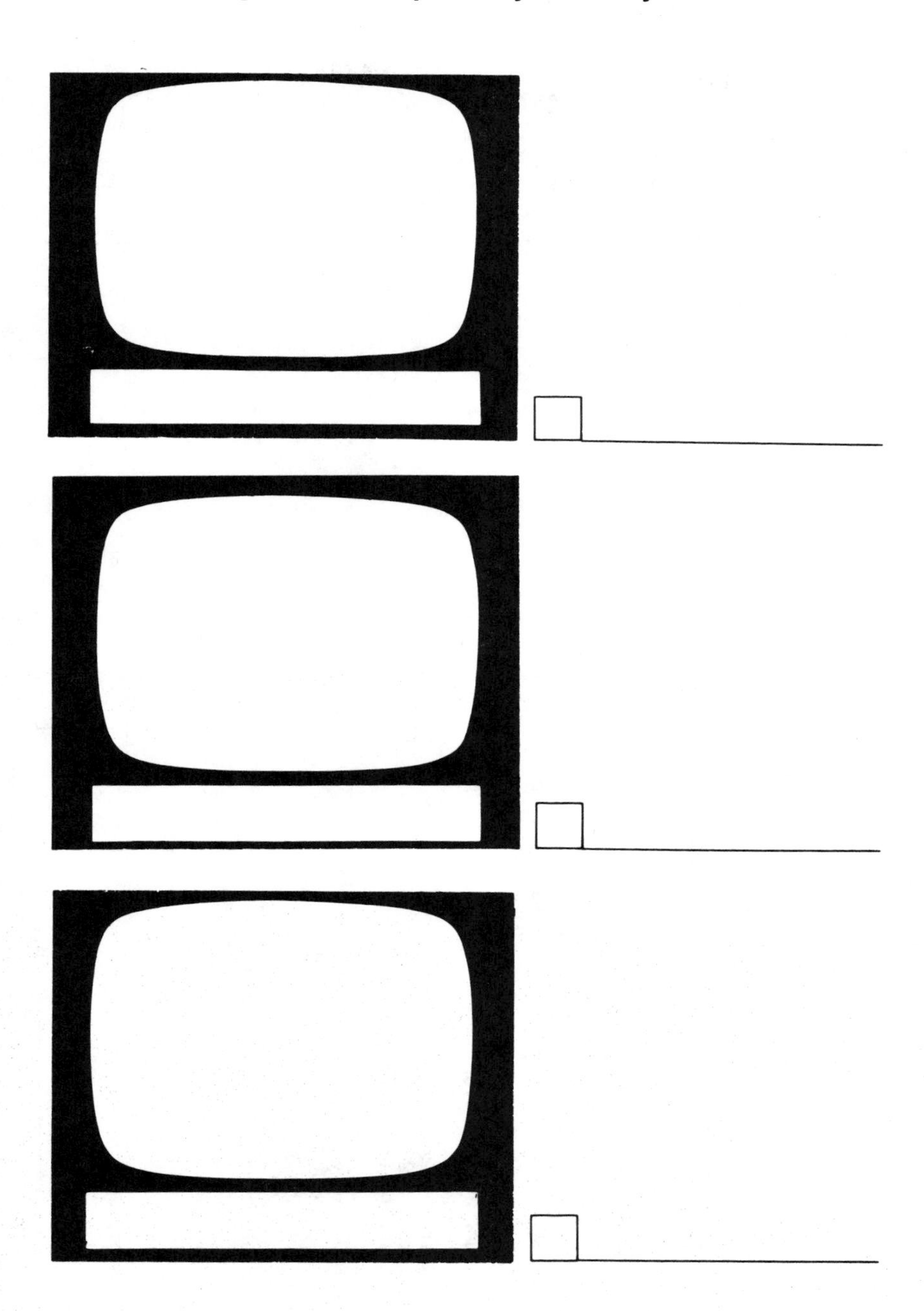

Figure 6.3: Graphic Arts Services Form

Job No.	Date In	Art Due	Project Due	Date Completed

☐ Design/General
Artist/Designer

Special Instructions/Information ____________________

Total Hours

Project Phases	Sched. Time (Hrs.)	Date	Date	Scheduled Revision			Comments
				Due	Due	Due	
Initial Design & Layout							
Client Review							
Design Revision/Comprehensive							
Typography							
Illustration							
Photography							
Photostats							
Keyline/Assembly							
Approvals							
Production Specs.							
Totals							

Figure 6.4: Sample Authorization and Release Form

For value received, I give and grant to Company X, its affiliates, their successors and assigns and any person acting under its permission and authority (hereinafter called "company"), the unqualified right, privilege, and permission to reproduce in every manner or form, publish and circulate videotapes, films, photographs and transparencies of me and my property, and recordings of my voice, arising out of ________________________________

____

____

and I hereby grant, assign and transfer to Company X all my rights and interest therein.

I specifically authorize and empower Company X to cause any such videotapes, films, photographs, and transparencies of me and my property, and recordings of my voice, to be copyrighted or in any other manner to be legally registered in the name of Company X.

I, for myself, my heirs, executors, administrators and assigns, hereby remise, release and discharge Company X for and from any and all claims of any kind whatsoever on account of the use of such videotapes, films, photographs, and transparencies of me and my property, and recordings of my voice, including but not limited to any and all claims for damages for libel, slander and invasion of the right of privacy.

I am of lawful age and of sound mind, and have read and understand this Authorization and Release.

Signed this ________of______,

19 _____.

____
Name

WITNESS:
____

____
Address

____

tween. The rationale for creating feedback mechanisms for media projects is for media production management to develop more control over the perceived success or failure of a particular project. Some organizations have developed elaborate feedback forms, while others rely simply on verbal (informal) feedback from clients. A feedback form for a videodisc, video, film or slide program might consist of the following items:

Title of program
Date
Size of audience
Kind of audience (internal, external, professional, hourly)
Technical problems (with the program or playback equipment)
Statement of objectives
A general question regarding the response compared to the program's objectives
A question regarding the technical quality of the program
An open-ended question, e.g., "further comments"
Respondent's location or organization

In addition, the name and address of the person to receive the feedback questionnaire should be prominent.

Administrative Forms

Forms which fall into this category include:

Audiovisual equipment services
Equipment trouble report
Equipment repair/maintenance records
Maintenance and inspection records
Equipment operating logs
Inventory logs (stock)
Inventory logs (audiovisual equipment)
Media library requests

Figures 6.5 and 6.6 give examples of two administrative forms.

INVOICES AND EXPENSE REPORTS

Since every organization will at some time use external resources to get a job done (e.g., scriptwriters, consultants, lighting designers, etc.) a system needs to be developed to handle the accompanying bills and

Figure 6.5: Maintenance or Inspection Record

DATE	SERVICE PERFORMED	TIME	PARTS USED	PRICE

Figure 6.6: Audiovisual Hardware Monthly Usage

____
Month Year

Hardware Type	Number of Requests	Average Loan Period	Total Hours	Comment

paperwork. Regardless of the amount of work to be performed by an outside vendor, the corporate user should require the vendor to forward a formal statement of estimated charges and a statement of what he or she expects to do for the organization. This letter then becomes the estimate of charges. Similarly, the organization could send the vendor a statement of estimated charges along with a statement of the work it expects to be performed. This kind of documentation can then be forwarded to accounting. When bills for the work begin to come in, the invoices can be checked against the statement of estimated charges. The statement of charges also provides the user with a check against estimates of time and additional charges.

Before the invoice is forwarded to accounting, however, the individual who contracted for the outside service should first check the invoice to make sure the service billed has been performed. This person should then initial and date the invoice and forward it to the appropriate manager for further approval. In some organizations, systems have been developed whereby the client gives budget approval authorization to media production managers. This system authorizes the managers to approve invoices from outside vendors, as well as other bills, including internally generated charges, up to the limit of the approved budget. Therefore, once an invoice from an outside vendor has been approved, the manager can forward the invoice to accounting for payment, thereby avoiding the need for the client to see all the bills and become involved in a lot of paperwork.

There are some limits to this type of system, however, and these will depend on how each organization has set up its accounting checks and balances. In one organization, media production managers may have invoice approval authorization limits of $1000, while in another organization the limit might be $5000. Such limits may depend on what level the manager is at in the organization. In any event, the unit manager or, at the very least, the media manager, should negotiate the most flexible authorization limits, depending on the normal course of business. If 90% of all invoices for external services fall between $500 and $10,000, an authorization limit of $10,000 would seem sensible.

Before bills are forwarded to accounting or to the client for approval, copies should be filed in two places—in the project file and in a master invoice file. Depending on the volume of outside services, it might be wise to develop the master invoice file by vendor.

FILING SYSTEMS

Media production generates paperwork that reflects the history and cost of the project. A filing system therefore needs to be developed. Each organization will be different, of course, but no matter what form the filing system takes, it should at least provide easy access by management to any initiated, ongoing or completed project. The system should also provide a basis for department management to develop cost analyses of the various aspects of the operation.

THE RATE CARD

As was mentioned in Chapter 2, after one or two years of experience, the media production department should begin to develop a rate card for its services. By charging clients for media production services, the department can prove that it provides services at or near break-even or, failing this kind of success, that media production services should be modified in some way to make the department more cost-effective.

Rate cards should be developed for each major service offered in the department. In "Administration," charges for the following services should be determined:

- conference room setups
- equipment loan-out
- media distribution
- library loan-out
- media presentations

The graphic arts unit could consider charges for these services:

- graphics design
- mechanicals

The film unit could charge for the following:

- slide design
- slide production
- still photography (studio)
- still photography (on location)
- cinematography (director of photography)

The electronic media unit could charge for these services:

- pre-production (surveys, script supervision, travel time, engineering preparation)
- production
- post-production
- duplication (if in-house resources exist)

The charge-back items represent the labor expended by in-house personnel for media production work. The total cost of production will depend on the amount of labor expended, plus out-of-pocket expenses (scriptwriters, processing labs or printers, for example), in addition to so-called per-diem expenses, such as ground and air travel, food and lodging and other miscellaneous expenses.

The rate of labor charges for each activity outlined above will depend on many factors, the most important of which is what it would cost the clients to go outside for these services. If, for instance, clients can obtain a film at $3000 per finished minute outside, and it would cost the client $3500 a finished minute on average to produce the film in-house, the organization is better off using external resources (unless other forces, such as confidentiality or proprietorship preclude such activity). The rate of labor charges will also depend on the volume of work expected for the year. Given the volume of work and the amount of charge-back for labor, will this combination result in a profitable, break-even or unprofitable media operation at the end of the fiscal year?

The final consideration is what the market will bear. There is no reason why an organizational media production department should not charge as much as it can for its services, while remaining cost-effective. For one thing, the higher the charges, the better the chances that a break-even operation will be realized. Second, higher charges seem to have a beneficial psychological impact on clients. Perhaps we can call this the Cadillac or Steinway syndrome: we expect to pay for quality. My experience has been that low charges for media production services sometimes develop an attitude of low respect for the in-house operation.

Higher charges can also have an impact on the media production staff, giving them a sense of partaking in and contributing to a business. The service they provide brings money into the business. There is something psychologically invigorating knowing that the service provided to a client is going to make a difference to the survival of the unit.

COST ANALYSIS

A good filing system, good records and thoroughly filled out project forms can help facilitate the development of cost analysis. Each major unit of the media production department should be able to provide the manager with at least quarterly, if not monthly, reports on the cost of doing business.

For instance, in the electronic production group, the manager should know the average costs of pre-production, production, post-production, duplication and distribution for video programs and develop an overall average cost. The manager of the film production unit should know the average cost of slide production, the unit cost of still photography and the average length of time spent by each photographer on projects. The graphic arts manager should know the average cost per graphics project, and so on.

Each organization must develop whatever cost analysis it needs to satisfy its cost-justification program. Cost analysis will prove invaluable from many points of view, one of which is the ability to answer the question, usually posed by novice clients or skeptical management, "Yes, but how much does it cost?" Another reason is that the cost analysis will give media management a control on where costs are increasing or decreasing so that adjustments can be made.

Finally, knowing the costs just makes good management sense. It provides another way in which media management can let higher management know that the production department is being properly managed. Knowing "the numbers" also provides a basis for giving clients an idea of the scope of a project. For example, a client new to the use of videotape or slides will appreciate learning that, "The average length of our programming is 15 minutes," or "The average number of slides in a one-projector program is usually around 65," and so on. Knowledge of the costs will also give producers, photographers and graphic artists a working idea of how long it should take to complete a particular project, which they can communicate to the client.

REPORTS TO HIGHER MANAGEMENT

Media production management should not wait to be asked "How are things going?" by top management. At regular intervals, media managers should report on the state of the department. For example, monthly reports can be sent to higher management regarding the volume and kind of work being performed, perhaps mentioning the following:

- the number of multi-image programs, videotapes, films, slides, photography assignments and graphics projects produced
- changes in personnel
- additions to hardware
- awards or special commendations received from clients
- out of the ordinary projects or assignments
- special training for media production staffers
- changes in procedures that have improved overall operations
- visits from peers from other organizations

Reports such as these keep the department in the forefront. They present an aggressive posture that can translate into higher management's perception that the department is being well managed. A periodic report also reminds higher management that the department is making a contribution to the organization.

The report should occasionally make comparisons such as: how many slide projects were completed in January of this year as compared to last; has the overall volume of photography assignments increased or decreased in the last six months; has there been a shift in the use of videotape from one application to another? Information of this kind can also be used to build a case for future requests. For example, if the volume of video production work is increasing, or the film unit is experiencing an increase in the volume of slide production, the next budget cycle might find media management requesting additional personnel, hardware or operating dollars. The monthly reports can tell higher management that change is taking place and when the request for additional personnel, equipment and operating dollars is formally made, it will not come as a total surprise. Periodic reports will have helped to build up the perception that the requests have a basis in fact, rather than in creative production fancy.

QUALITY CONTROL

Quality control checkpoints should be developed for each aspect of the media production operation. The manager of each unit should be responsible for developing quality control mechanisms. For example, a system can be developed to check on the technical quality of videocassettes before they are shown or distributed to field locations. Even the labels on the videocassettes should be double-checked to make sure they contain the right program name, length, release date and correct spelling of the names of any executives. Similar quality control checks should be exercised on slides, photographs, films and so on. Regardless of the medium involved, some system should be developed so that someone other than the originator of the project has a look at the output to check for technical quality.

THE OPERATIONS MANUAL

Most of the media production department's activities should be formalized into an operations manual. The operations manual will provide a new manager with a ready reference as to how things get done, what procedures are used, what policies are followed, and so on. It also serves as a handy reference for anyone working in the department.

At a minimum, the operations manual should contain the following information:

- A description of the department's purpose and objectives
- A specific breakdown of the purpose and objectives of each unit within the department
- A formal description of all jobs in the operation, including general statements, requirements for the position, job performance expectations and chain of command
- Measures of performance standards—means by which management can evaluate the performance standards set forth in the job descriptions
- The department's rate card
- A formal organizational chart
- Pertinent organizational policies, such as those regarding vacations, sick leave, holidays, EEO requirements, etc.
- Departmental policies
- Production procedures—each unit manager may have developed procedures that everyone in the department must follow which should be outlined formally in the manual
- Forms—all the forms used in the department should be in the manual with an explanation of their use and function
- Lists of frequently used outside vendors, any specific arrangements with these vendors and copies of contracts
- Formal procedures for using outside vendors
- Statements regarding the use of copyrighted materials

LIBRARIES AND CATALOGS

In time, a media production department may develop a sufficient number of graphics, photographic prints, slides, 16mm films, audiotape programs, multi-image programs, videotape programs and even perhaps videodisc programs to justify the creation of a library. A media library can serve various purposes. It serves as an archival depository for at least one copy of each program the department produces. It makes it possible to retrieve programs to be used again, even years later.

A media library, then, implies a media catalog. Such catalogs are of enormous value, especially if there is a lot of material, and if many different locations within the organization need to use the library. Catalogs could be created for each of the media departments; moreover, each department catalog could be subdivided by application. For example, a slide library (sometimes called a slibrary) catalog could be divided by communication areas: internal communications and external communications.

7

Using Computers

A media production operation that is not using computer devices for running its day-to-day operations, or for some aspect of media production, is not taking full advantage of its potential role.

For example, the computer allows people to create scripts in various incarnations for client review in a shorter time frame. It can generate expense information in different forms and more quickly than previously possible. The computer allows for the generation of "more variations" of information, just as the word processor has made the process of revision faster and the presentation of the material more professional.

Professionalism is what the computer adds to the managing of the media operation. Most other departments in the organization are using computers in one way or another, and to remain professionally competitive (and administratively adept) the media production department must use computers effectively.[1]

WHAT THE COMPUTER HAS TO OFFER

Personal computers can be used for a variety of media management and production functions, namely:

- Departmental budgeting
- Individual project budgeting
- Department accounting

- Job cost tracking
- Tape libraries
- Edit logs/time code logging off
- Word processing
- Electronic mail
- Proposals
- Online research

Probably the most frequently used personal computer devices in media are the Radio Shack TRS-100, the IBM PC XT and AT, and, to a lesser extent, the Apple Macintosh.

There are various reasons why a media production department can take advantage of the features of computers. According to Andy Bobrow, executive producer of BioMedia, a New York-based medical and scientific video production company, "The personal computer allows a media department or small production company to increase productivity. A personal computer system can allow a group of four people to do what it might have taken 24 people to do in the past. Clearly, increased productivity is the single major benefit."

Bobrow points out that the personal computer:

- decreases the amount of time needed for script rewrites
- allows the user access to databases and speeds up research time
- decreases the amount of time needed for management functions

The computer thus increases efficiency.

Bobrow indicates that at large corporate media operations, functions like budgeting and accounting are probably standardized according to corporate-wide guidelines. But he observes that computers have created a democratization process. Depending on the needs of the department, the personal computer allows a user to fit the capabilities of the computer to his/her individual needs.

Jim Brady, former manager, video communications, Union Carbide Corp., and now head of the Connecticut-based Concept Associates, observes that "Personal computers provide users with speed. What used to take someone three to four hours to do the old way now takes a fraction of that time. You do spend a lot of time setting up and creating a template, for example, for the Lotus 1-2-3, but in the long run the computer gives the user a time edge."

Brady also comments that "The computer helps you do your job better. Anyone who has to write or create should be using a personal computer. It makes a difference. You can be more creative. You can move things around. You can do things on the computer in half the time they would take with a typewriter. The brain is just too fast for the typewriter. I know for myself, with the computer I can go as fast as I can think.''

GETTING INTO COMPUTERS

According to various users, the basic investment in a computer hardware system will vary from $2000-$3000 to $50,000 for a system that includes at least one graphics station.

Getting into a computer system for the media production function involves steps very similar to those outlined at the beginning of this book. The first question is: what do we do here? In other words, what do we do of a fairly repetitive nature that could use the speed, flexibility and analysis powers of a computer?

The second step is to analyze these functions; e.g., budgeting, scripting, media project cost tracking. The third step is to look for areas where efficiency can be improved.

Once these needs have been established, the next step is to research the computer hardware/software market to identify and select a system that will serve these needs.

There are several pitfalls potential users can encounter when getting into a computer system. These pitfalls are similar to those mistakes corporate media managers can make when selecting production equipment. Andy Bobrow points out, for example, ''Just because it can be done doesn't mean it should be done. People get hardware crazed, and that can be dangerous.''

He cautions, ''You have to systemize things *before* you computerize. The computer isn't going to do it for you. You must also make an allowance for the learning curve. Sometimes this will take two weeks, sometimes it will take six months. But you must allow time to learn and debug the system.''

Bobrow also points out that ''Sometimes it's better to do something the old-fashioned way, as in using a paper and pencil. Sometimes the computer takes more time and more work. You have to ask 'Is it going to be more or less work in the long run? Do a lot of people need access to the information? Is there sufficient volume of information (as in a tape library) to warrant the use of a computer?' ''

Brady and other users point out that a hard disk drive, as opposed to a floppy disk system or a two floppy disk system, provides access time that is almost instantaneous. The printer is also an important aspect of the system. Brady comments, ''At the very least you should have a dot matrix printer, although most people want laser printer quality. You can get a daisy wheel printer, but they seem to be going out of style. In some cases a department might have a laser printer on a shared basis, with people using dot matrix printers on an individual basis for rough drafts.''

Many users have found that computers are not generally user friendly. And some software manuals are better than others. It helps to have some sort of mentor, whether an associate, a class or a friend. Brady states, ''It's easy to get discouraged. There is a learning curve. But once you get past the first learning hurdles, it can be a very productive experience.''

HOW MEDIA DEPARTMENTS USE COMPUTERS

GTE

The GTE Co. has taken its use of personal computers and software to a fairly high level. According to Billy Bowles, director/visual communications, the departmental computer system is being used to manage a diversity of media department and production functions and several media department operations.

Says Bowles, ''We have developed what we call the MMS—Media Management System. We started to develop this early on, in 1984, and now it does more for us, frankly, than we really need.''

The MMS, using the IBM-PC and Dbase III, handles a variety of functions. For example, it tracks media projects. By inserting various codes, media department management and production personnel can scan more than 150 line items day by day, whether the project is in pre-production, production, post-production, on hold or completed. Bowles comments, ''The computer system lets us account for everything, and it gives us great flexibility. From a logistical point of view, the system gives everyone up-to-the-minute information. We can tell if we are over or under budget. It also provides the client with specific information during the budget approval process and the production process.''

The system also lets management track the productivity of individual producers. Management can gauge, for example, the average length of time it takes to produce certain kinds of programs.

Bowles' operation has organized programs into four categories: Classes I-IV. Class I identifies a program that uses the talking head as its primary production style. Class II identifies a program that is ''Intermediate'' in production style. Class III is a program that is ''Advanced'' in production style. And, Class IV is a program that is highly complex.

The MMS allows media production staff to determine such things as cost per minute and average length of time needed to produce a particular class of program. The system can also track the usage of slide and multi-image presentations, for example, number of slides per presentation. Types of photographs are also tracked by the system.

In addition to tracking media production projects, the MMS is becoming an important management tool for GTE on a national basis. Because the telecommunications division of GTE is consolidating and reorganizing, the various media production departments that service the telecommunications division around the country are responding to these organizational changes.

According to Bowles, GTE has (as of this writing) 13 media centers. With the help of the MMS "...in about two years these 13 centers will be consolidated to four." According to Bowles, with the help of the Media Management System, headquarters media production management has been able to get a handle on production title duplication. Bowles foresees that duplication could be reduced by 30%. He indicated that in 1988 the telecommunications division media department alone produced 648 programs, of which there is a perceptible duplication of effort.

Bowles believes the MMS overall has provided 99% more accuracy in tracking the costs of various media production projects, as well as the overall cost of the department. In time, the MMS will help manage various media production departments for GTE as a national service to the company's telecommunications division.

SAS Institute

The SAS Institute, a Cary, North Carolina-based computer software company, makes extensive use of computers for media department and project management.

According to Bill Marriott, manager, corporate video, the company has been involved in the use of computers since about 1982. (The use is a direct outgrowth of the company's own product line. SAS develops and markets computer software for data entry and retrieval, word processing and graphics and operations research for mainframe, mini- and microcomputer applications.)

In 1982 the company started producing video programs to accompany its computer software and workbooks, creating a video/software/workbook training package. The company had previously marketed its product primarily through telemarketing.

As if in reciprocation, the video production department started to use the company's own product to help manage the department and individual projects. Says Marriott, "The use of the computer software was and is an effective way of scheduling projects and scheduling the development of a particular project. Sometimes we handle projects that last up to a year from start to finish and involve eight different departments. The computer software allows us to use a Critical Path Method for developing the schedule.''

Marriott indicated that the application of the computer software had an ancillary effect: ''The more we used the computer software, the more we got organized. Since we are working in a company that is in the computer software business, our embracing of the product showed the company that 'we had our act together.' And that made a very favorable impression on management. As a result, today we are very involved in the company's development of new products. Because we've gotten to know our company's product very well, we've expanded our role as a department.''

Marriott's department uses the computer software for a variety of applications:

- word processing and scripting
- talent contract revisions
- talent file (each file has a detailed description of the talent's audition, characteristics; these details are cross-referenced with a variety of factors, including demographic profiles)
- stock footage file (daytime/nighttime shots, location, from ground/from the air, on a worldwide basis)
- spreadsheet analysis for the department
- spreadsheet analysis for individual projects
- equipment maintenance (every morning the engineers check this file for equipment problems; the file provides a complete history of the equipment, a daily review capacity and prioritizes what equipment needs fixing)

Marriott's department produced approximately 120 projects in 1988, ranging from short two to three minute public relations pieces to 90-minute video courses. Many of these programs are distributed in several languages to SAS's 1500 employees worldwide.

Marriott says the computer system provides information ''at our fingertips. It gives us an organizational ability that we didn't have before. Everybody has access to the information. And there are also fewer memos floating around. We can present a crisp, businesslike stance to our company's management. It helps make us successful.''

Wausau Insurance Companies

Wausau Insurance Companies (Wausau, WI) uses a computer software-based system to manage the distribution of several thousand titles to its customers. Wausau provides insurance to many manufacturing companies. The purpose of its media library is to provide customers with current information on safety procedures.

According to Doug Lay, manager, media and reference services, Wausau distributes media materials on video, slides and 16mm film to its customers on a request basis. About 75% of the library concerns safety procedures. The company distributes titles to about one thousand existing customers. Lay's department receives several thousand requests annually.

Lay's department has installed a computer-based system using software developed by the Tek Data Systems Co. (Libertyville, IL). It is called the Automated Media Management (AMI) System and runs on IBM/PC AT hardware.

The AMI System consists of one or more microcomputers, a printer, one or more videodisplay monitors and keyboards or terminals, and an integrated set of library management computer programs. The system operator communicates with the computer via the terminal, which instantaneously displays system-generated messages to lead the operator through the various library management operations.

Automated functions performed by the standard AMI System include:

- Data maintenance procedures to allow immediate updating of information on borrowers' accounts and items
- Scheduling procedures to allow easy definition of comprehensive holiday and shipping schedules for accounts, shippers and the equipment library
- Interactive booking procedures to allow immediate and future media reservation requests to be entered and confirmed online
- Daily checkout, check-in, and confirmation procedures, which include the printing of shipping labels, shipping lists, call lists, due lists and confirmation forms
- Statistical tallying to record bookings, turndowns and shipments by account and item
- Printed reports of the database contents and usage statistics
- Backup and recovery procedures to minimize the loss of data in the case of equipment failure.

Prior to the installation of the computer-based system, Lay's department used an "automated card catalogue, essentially a system that was print/manually based."

Since the installation of the computer system, the department has derived a variety of benefits. According to Lay, "Essentially our system is a booking system. Customers call in, fax in or write in requesting a particular title or general topic. The manual system was rather involved. With the computer-based system we can give our customers much faster response. All the information is right there on the screen. The customer service person doesn't have to spend hours trying to find a title, or the availability of copies of a particular title. With the AMI we've been able to improve customer service without additions to staff. We can also get better statistics out of the system.''

Another benefit of the system is that it has precipitated a consolidation of media distribution activities. ''Several parts of the company distribute media to our customers. With the AMI system, we are beginning to do a better job of tracking what is where and how much we have of what title.''

Lay's library consists of 40% videotapes, 40% 16mm, 15% slides and 5% audiocassette and PC-based training.

Learning the system was also a positive experience. While it might take someone several weeks to learn the manual system, the computer system, according to Lay, could be learned in a few days: ''We expect a shorter learning curve for someone using the computer-based system for the first time.''

McGraw-Hill

McGraw-Hill Inc. (New York, NY) is also using the AMI System. According to Jim D'Elia, manager, training systems, customer service and data entry, McGraw-Hill has a rental library of around 3000 titles, of which there are 50,000 units.

McGraw-Hill formerly used a manual card file system, which was administered by seven or eight people. The AMI System was installed in June 1987. Administration of the rental library now requires two or three people.

McGraw-Hill is also finding that the computer-based software package enhances customer service: ''While our clerk is talking on the phone with a customer, she has all the information right in front of her. She can make changes right on the screen. She can look up the title, the rental history. The system can even produce the shipping documents.''

The learning curve has also been reduced. The manual card system had taken two to three weeks and more to learn. The Tek Data representative spent a week training D'Elia's staff. According to D'Elia, the staff was up on the system within two to three weeks after that.

Air Products

Air Products (Allentown, PA) also makes extensive use of computers for media production management.

The Audiovisual Department, headed by Manager Brian Sullivan, uses Lotus 1-2-3 software for media production management and Power Base software for engineering management. With an IBM-compatible system, Sullivan's department uses the software for budgetary purposes, such as tracking how much an individual producer spends on particular kinds of projects and how much a client spends on a project. The software also stores equipment records, wiring diagrams and equipment layouts.

The department uses a separate computer system for word processing, to produce proposals and scripts, for example.

In an emerging application for the company, Air Products also uses the computer system for scheduling one-way teleconferences from the Environmental Protection Agency, which provides masters degree courses in electrical engineering. The video courses are downlinked to a TVRO (television receive only) dish and then redistributed to various conference rooms through a local area network (LAN) system. Sullivan's computer system tracks the scheduling of the video courses and the use of the TVRO and the LAN.

His department also produces slides and overheads. In effect, the department's computer software tracks about 60-70 projects a year.

Says Sullivan, "The computer software is definitely a timesaver. We're able to play a lot of 'what-if' scenarios. For example, we can ask the computer what will happen if the client reduces the budget overall, or cuts a particular segment of the budget. The computer software gives us a menu for each part of the production and allows us to respond definitively to our clients.

"We're a full charge-back department. The computer software makes us a lot more organized. Handwritten material is only used as backup.

"We're still going through our learning curve, even though we've been at this for about two years now. Every time I sit down at the keyboard, I find new things we can do with the system. We're always picking the brains of computer department people. Sooner or later we're going to get up to mainframe applications. There are applications we probably haven't even thought of yet.

"Overall, though, the use of PCs has enhanced our image. We are perceived today as being more up-to-date administratively because we're using computers."

Georgia-Pacific

Georgia-Pacific (Atlanta, GA) developed and installed a customized computer system.

Designed by Xymox System, Inc. (Roseda, CA), the software package runs on an alpha micro mainframe (i.e., mini mainframe), and provides the Georgia-Pacific television and photography department with several production-oriented functions:

- daily procedures
- activity reports
- videotape library
- custom reports
- system management

The computer system also provides the department with the ability to handle accounts payable, accounts receivable, purchase orders and word processing.

"The major reason for the development of the system was volume," says Don Blank, director of television and photography operations at Georgia-Pacific. Blank's operation not only produces videos for the corporation; it also provides television and photography and slide production functions for clients outside the company. In fact, 70% of its work is for outside clients.

"We're a profit center. In this capacity we're doing about 50 corporate videos for Georgia-Pacific and about 55 corporate programs for outside clients. On top of that we do about 80 commercials a year, most of them regional.

"We opened in 1983 and we were in the right place at the right time with, apparently, the right kind of approach. Because of the volume and the kind of extensive facilities we have, we needed a computer software system to help us manage our business."

The system took about eight months to develop once the organization decided on a vendor. Prior to its installation in September, 1988, the software was tested at the Post Group in Los Angeles for six weeks.

Blank points out, "We provide a turnkey operation to many of our clients. That means we create a tremendous volume of paperwork. And most of it is handled by creatives—engineers, directors, lighting directors, guys in blue jeans and sneakers. We don't have an adequate clerical staff. As a result, there was a high error rate.

"With the system we're realizing cost savings. The system has become a money-maker. It's increasing profits by eliminating missed billings. The major benefit is that it is creating consistency throughout our operation.

"The major problem," says Blank, "was simplifying the administrative system. Everybody does paperwork. The system is self-prompting, so that the least experienced individual can get up to speed in a few days. We expect the system will have paid for itself in about two years."

Blank points to other advantages. "It has great flexibility. We can add items and get more terminals without calling the software designer. There's ease of information dissemination. People are better informed. We've eliminated equipment conflicts. There's better input. Everybody knows what's going on. We're creating more accurate bids. And there's been a tremendous cut-down on paperwork."

Blank indicates that 70% of the work is video-based, with the balance split between still photography (15%) and slides (15%). Blank explains that while the system is primarily used for video, the software can be adapted with ease to any of the other media production services his department provides.

CONCLUSION

Persuading management to budget funds for the purchase of a computer system is a process similar to that outlined in this book's first chapter.

The first step is make sure the need for a system is clear. As Bobrow pointed out earlier in this chapter, "Just because it can be done by computer, doesn't mean that it should be done by computer."

It is probably true that most corporate media production departments could use a computer system for some aspect of their production operations. If the need exists and it is made clear to management, the next step is to thoroughly research the possibilities and make an intelligent choice among computer alternatives.

Most important, media managers should look for every cost advantage the computer system can provide. The media manager should use an effective sales technique in his/her presentation to management: presenting the features of the computer system and the attendant benefits side by side, especially those that translate into cost savings and/or employee productivity.

Once purchased, patience should prevail. Allow time for the system to be installed, tested, receive input, be debugged and then be tested again.

Over time, the media manager should continue to explore new applications with the existing system, especially new software packages.

In sum, the successful introduction of a computer system into a media production department can only enhance the long-term viability of the department.

A Word About Software Packages

Various software packages used by media managers and producers have become popular, including Lotus 1-2-3 for accounting purposes and WordPerfect, Displaywrite, Microsoft Word and MultiMate for full-featured word processing.

Comprehensive Video Inc. of Northvale, NJ, has developed several software packages for the electronic media (video) manager/producer-director. They include ScriptMaster™, a two-column word processor for scriptwriters; SpellMaster™, a dictionary that includes film and video production terminology; Library Master™ for cataloging tapes; Budget Master™ for budgeting and tracking, Tape Master™ for tape logging and several packages for editing.

FOOTNOTE

1. For an excellent volume on the use of computers for video production, see *Computers in Video Production*, by Lon McQuillin, Knowledge Industry Publications, Inc., 1986.

8

Marketing the Media Production Department

No product or service gets used until someone sells it.

This has been true of every product ever developed, either by design or serendipity. This has been true of virtually every communications medium described in this book. When the Greek alphabet was developed about 700 B.C., there were those who decried it; they felt that people would lose their memories. When printing developed in about 1455, there were those who decried the changes it brought; knowledge moved faster from place to place than before. When television developed, there were those who decried its use to televise baseball games; they thought no one would come to watch.

Today, the corporate media department faces an ongoing challenge: developing and keeping clients. Today's corporate media production functions range in size from one-person operations using a combination of in-house equipment and external resources to multimillion-dollar operations that include in-house production and technical personnel, studios, field video production equipment, still photography labs, multi-media production facilities and teleconferencing rooms.

Given the constant fluctuation of external economic changes and the constant internal pressures from management to increase productivity and justify capital expenditures, corporate media production functions have to be economically viable.

EXTERNAL CLIENTS

To this end, several in-house video production departments have not only aggressively sought internal clients, but also have looked to the "outside" for clientele. Table 8.1 lists more than 20 companies currently using their in-house video facilities for outside clients. Several conclusions may be drawn from this list. First, these are video production departments and therefore do not reflect the external client use of slide or photographic services a media department could offer.

Second, the general consensus among the experts interviewed for this section was that, at best, perhaps 5% of the companies in the United States with in-house television production operations have gone "external" with their video operations.

Third, the video operations ranged in size from a capital investment of $250,000 to several million dollars. Further, the proportion of external to internal clients ranged from 10% to 100% (with the parent corporation as a client).

Last, and perhaps most important, the primary motivation for the "external" client perspective was "expense recovery" followed by "profit." While there were some exceptions, this illustrates the "expense pressure" put on corporate video departments and is indicative of a significant change in the economic environment in which media/video departments now operate.

In the 1970s, there was a strong trend towards bringing video production operations in-house as a cost-effectiveness measure. In retrospect, however, this may not have been such a good idea. Many companies have reevaluated the cost-effectiveness of their video operation and decided that, while convenient, the capital and operational cost of an in-house operation may not be viable. In fact, several video/media operations have been shut down, even though they were cost-effective, because the corporation decided it was not in the video business.

This contrasts sharply with the continued existence of internal slide-making and graphic arts operations. The major difference between the electronic and graphic/film media is the capital investment required to remain up-to-date. There have been technological changes in the way slides are made (via computer) and the way print pieces are made (via desktop publishing). But the capital investment, plus the cost of employee salaries and benefits, of a moderate-to-large-sized internal video operation is far more extensive. It is also very visible.

This is why some in-house video departments have sought outside clients. The revenues derived from outside clients help defray the cost of the internal department. The revenues help the department remain

Table 8.1: Inhouse Facilities with Outside Clients

Parent Company	Facility	Location
Anheuser-Busch Companies, Inc.	Busch Creative Services Corp. Innervision Production Inc. Optimus Inc.	St. Louis, MO St. Louis, MO Chicago, IL
AT&T	CSO Video Resource	Piscadaway, NJ
Contel Corp.	Contel Creative Services	Atlanta, GA
Dayton-Hudson Corp.	Studio 11	Minneapolis, MN
Drexel Burnham Lambert	Broad Street Productions	New York, NY
Empire of America Federal Savings Bank	Sherwin-Greenberg Productions, Inc.	Buffalo, NY
The Equitable Life Assurance Society of the United States	A/V Production & Design Group	New York, NY
Georgia-Pacific Television	Georgia-Pacific	Atlanta, GA
Gulf Power	Vision Design Tele-productions	Pensacola, FL
Harcourt Brace Jovanovich	HBJ Video	Orlando, FL
Health East, Incorporated	Health East Communications	Allentown, PA
Industrial Bearing and Transmissions	IBT Video Productions	Miriam, KS
Marshall Field's	Image Field's	Chicago, IL
McDonnell Douglas	Video Department	Long Beach, CA
Payless Cashways	Take 2 Productions	Kansas City, MO
J.C. Penney Co., Inc.	J.C. Penney Communications	Dallas, TX
Prudential Insurance Company of America	The Prudential Audio & Visual Center	Newark, NJ
SAS Institute	Video Department	Cary, NC
Salt River Project	Audio-Visual Division	Phoenix, AZ

Table 8.1: Inhouse Facilities with Outside Clients (con't.)

Parent Company	Facility	Location
Union Carbide Corporation	West Ridge Productions	Danbury, CT
Western Atlas	Corporate Communications	Houston, TX
Weyerhaeuser Co.	Weyerhaeuser Video	Seattle, WA

technologically up-to-date. And, external clients help smooth out facility utilization peaks and valleys.

The companies that are promoting themselves to outside clients are using a variety of techniques, including word of mouth, newspaper advertising, direct mail, rate cards, press releases, presentation reels and telemarketing.

Several of the internal video production operations are using a new name to separate themselves organizationally from the parent company. This is true of Take 2 (Payless), Broad Street Productions (Drexel) and West Ridge Productions (Union Carbide). But there are pitfalls to being a resource to outside clients. Don Blank, who runs the highly successful Georgia-Pacific television operation, is quick to point out that his internal clients come first. Says Blank: "If your operation is having a hard time cost-justifying itself to internal clients, trying to get outside clients to make up the difference is only going to make things worse. If you have a strong operation vis-à-vis your inside clients, then going to the outside for clients is building on strength."

Blank emphasizes that his internal clients are the first to be served; after all, the company provided the initial funds to start the operation. He also points out that it was the department's intention from the beginning to seek outside clients. (The reason was simply a response to market demand.The facility was built at a time when the broadcast quality operation was the only one in its market.)

Clearly, though, most organizations do not have moderate-to-large video/media production operations. Nor should every company with a large facility service outside clients; they might be too busy servicing their internal departments.

But all media departments must employ some form of marketing (whether to internal and/or external clients), in order to remain economically viable. In time, the department that does not market itself may find itself out of existence.

To reiterate Blank's conclusion, a corporate media department, let alone video production operation, must first successfully market itself to internal clients to remain viable in the long term.

MARKETING TO INTERNAL CLIENTS

Publicity and Promotion Techniques

The media production department that does a poor job of marketing itself within the organization will have little credibility with organizational clients seeking to enhance their own communications efforts. In addition, the corporate media production department serves a function similar to that of external agencies. As an "in-house agency," the department must act to get the organization's attention, and to "sell" its services. Publicity and sales promotion are two ways of accomplishing this.

There are various ways in which an in-house media production department can get publicity and promote itself. One obvious way is to get written up in the organization's house organ. A suggested hook for the feature could be "How does an in-house media operation serve an organization?" Or it could be a description of a special project. The house organ can also focus on each unit of the department, which means that over a period of time each aspect of the department can be covered in some detail. The house organ is a good place to publicize recently completed projects, or promote existing resources, such as a listing of tapes in the video library, or the department's slide library. Because the media production department's output is highly visual, photographs of the operation or of a project in the process of production are a natural. Photos could show the head of the organization making a videotape or a film, how a multi-image production is done, or the development of a storyboard for a slide program. Media production is graphic; it lends itself beautifully to graphic promotion.

The house organ is just one natural publicity outlet in the organizational system. There are others. Bulletin boards, cafeteria spaces, even office elevators can all be used for promotion. Media management can also use the facility itself as a showcase. Blank walls can be used to display production shots from various programs. The general access areas can also be used to display awards garnered by the department. These efforts should convey the message that successful and effective work is going on in the media department.

The media production department can also garner publicity and promotion indirectly by either soliciting, nurturing or actually writing features on the department for publication in the trade press. Once published, the article can be framed for display and copies can be forwarded to key management personnel. Depending on the organization, the article may be spotted without it having to be distributed. All the better. It is wise to touch base with upper management to let them know of the press' interest in the operation.

Sales promotion can be done by setting up meetings with key personnel or potential clients who are not familiar with the department's services. At these meetings, the manager can present clips from video or film programs, show a particularly effective slide presentation, give the client a tour of the facility, introduce production personnel, and so on.

Clients can also be marketed to by individual memo. For example, say the graphics group recently completed the design of a new logo for an existing product. A short memo to the marketing director of another operating division mentioning the new logo might get some attention and generate business.

An essential element of promotion is the development of show reels. Each unit should maintain an up-to-date promotion reel. Graphics arts should have a portfolio of recent and effective graphics; the film group should be prepared to show various types and styles of slide production work and still photography (both black-and-white and color); the electronic media group should have videocassettes showing various styles and applications. The department might develop an overall applications reel, or separate reels for marketing, training, employee communications, external communications and so on.

In all, publicity and sales promotion should have the long-term effect of generating business and keeping existing clients. Publicity and promotion can also help generate high employee morale by providing an opportunity for everybody in the unit to get credit for their contributions to the department. High morale helps improve operations, develop innovative output and keep the media production department in business.

Marketing and Selling Techniques

The media production department is in competition with all external media production companies. It must seek out clients, and prepare and submit competitive media project proposals. In sum, the in-house media department is a business. As such, the department must be marketed not just once a year when budgets are reviewed, but all the time. The media manager and the various unit managers must also be prepared to market and sell their "products," if the department is to remain viable.

It might seem odd that an in-house department should have to take on the marketing stance of an outside agency. Yet it seems that more and more organizations are moving toward decision package analysis techniques; in increasing numbers, organizations are taking the position that staff departments should put a price tag on their services and that these services should price their value by bringing in enough work to pay for themselves. In this kind of climate, media management must be prepared to adopt a strong marketing and selling attitude.

What follows are some techniques for bolstering the media department's status in the organization:

Know the organization. Successfully selling the media department requires talent, tenacity and good timing. It also requires self-confidence and knowledge—knowledge of clients' communications problems, of the organization's business, its products, customers, employees, and its past, present and future.

Know other organizations. Managers should know how other organizations, especially similar organizations, are using media—what equipment they use, how they are staffed, the kinds of projects they produce, how much each project costs, how heavily they rely on outside vendors, their failures and successes, how they evaluate their projects and how they sell management on media.

Develop media seminars. Show what can be done, let clients respond to examples and let them get involved in the decision-making process. Let them teach themselves how effective media can be. Bring in outside experts on media production and bring in managers and personnel from other organizations to tell their stories.

Produce quality work. Nothing is more damaging to a media department's image than a project stamped by its audience as amateur. And while the quality of a project does not necessarily ensure communications effectiveness, lack of it can instantly negate the real value of any content. While this does not mean that every media project has to cost a fortune, it does mean that shortchanging or shaving expenses from the cost of production can only result in a product that reads, sounds and/or looks like it was done on a shoestring budget. A media project should not only be cost-effective, but it should also reflect the quality of professionalism all clients expect.

Project a professional attitude. Clients want assurance that they are dealing with professionals. This means understanding the advantages (and disadvantages) of the various media, asking informed questions, being organized, following up, etc.

Get out of the office and promote. Take the initiative. Find out who could use the department the most. Which departments have the most need? Who in management is most disposed to the use of media? Ask for a 15-minute interview, and let it be the first of several.

Take advantage of chance. Make the unexpected work to advantage. A chance meeting in the corridor with a member of management who mentions he's having a communications problem could turn into an opportunity for a media project. Be alert to those opportunities.

Follow-up. A casual suggestion to a prospective client can be turned into a media project only if the client has something he/she can look at in black and white. Follow informal suggestions with a written proposal. Keep it short and to the point.

Learn to say no. This means taking charge. Let clients understand that if they want the right results, there is a proper way of getting them. Remember, the media production department wouldn't have clients if it weren't staffed with experts. If the client were the expert, in-house media professionals would not have jobs.

Demonstrate and educate. Show client A what was done for Client B (and vice versa). Show other clients what was done for clients A and B. And show all of them what other organizations are doing. A picture is still worth a thousand words.

Develop a mailing list. Who in the organization is involved in using media? The basic idea of a mailing list is to keep people informed. Make them feel part of an effective communications system.

Distribute pertinent literature. Forward copies of appropriate articles on media to select clients. Summarize the results of pertinent studies. Compare studies findings to the organization's use of media. Communicate this information.

Use innovation sparingly. Tried and true methods may be old hat to you, but to a new client this translates into: "I understand what you're suggesting." Use innovative techniques only when they contribute to solving the client's communications needs. To paraphrase the design philosophy of architect Frank Lloyd Wright, let the form and technique of the media project grow out of the client's needs and the project's contents.

Ask for written testimonials. If a media project is a special success, ask the client for a written testimonial. Make copies of it and send at least one copy to higher management in the form of an FYI. Keep a "testimonial" file and review it at least once every three months. The file will be a terrific confidence builder; it will also enhance the department's reputation with management.

Initiate the second sale. So you just finished a project for a new client. Is that the end of it? Perhaps that first project can turn into a series of projects, or a follow-up program three, six or even 12 months later. In short, always keep long-term interests in mind.

Cross-sell. Banks like to sell a checking account customer a savings account, or vice versa. Similarly, in the corporate media business, you can sell clients on various applications. Just because a client uses video for training purposes does not mean he/she could not also use it for management communications.

Use existing publicity outlets. Let's say you just finished producing the company's update of a five-year-old orientation program. If you took photos of the production in progress and shots of the program itself, get them published in the company house organ. Make posters using scenes from the program. Ask local employee relations personnel to display the posters.

Formally evaluate programs. Sooner or later management will want to know if the projects they paid for did what they were supposed to do (presuming that objectives were defined). Don't wait to be asked. Have an evaluation procedure designed before production starts. Summarize the results and communicate them to the client and other managers who could benefit from the feedback (e.g., using the mailing list you developed earlier).

Confront the stubborn ones. When you've developed a sufficient number of ongoing clients, and the number of projects has increased, confront potential clients who initially gave you a negative response. In effect, ask, "What do I have to do to get your business?" Sometimes the answer to this question will reveal a misconception—about various media, about the media department, about production costs, about access to the department's expertise. Use this opportunity to change a prospective client's attitude.

Track the department's progress. At least once every six months, list and summarize all management and production activities. Let management know how far the department has come. Not only does it serve to substantiate the department's growing reputation with immediate management, it also provides a quick picture of the extent of media use within the organization. One of the advantages of working in the organizational media business is that no one customer dominates the use of media. The department should have many clients from various parts of the organization, if media management has done a good job selling. So, while immediate management might perceive that the use of media is low (because they use media infrequently), it may in fact be extensive! A summary of programs and clients will help change this perception.

Be aggressive during economic hard times. There is a tendency in organizations during times of economic uncertainty for management to pull back and take a more conservative attitude toward plans, projects and policies. Whether the economic uncertainty is real or imagined, mild or severe, what counts is the "perception" of the uncertainty. Don't be put in the position of losing business when the organizational mood may be overreacting to the situation. Maintain an aggressive attitude, despite the impulse to become conservative along with everyone else.

Look for more business. Suddenly, department budgets for communications are reduced by 15%. In this economic environment the media production department cannot afford to "wait it out." If it stands still, it may find that its clientele has dried up completely. Managers should draw up an extensive, detailed list of potential user clients: marketing, training, safety, corporate communications, environmental affairs, consumer affairs, financial analyst relations, production, personnel, medical, quality control, etc. Put the list of potential customers in rank order. List as first those who would be the most likely candidates to use media and so on, down the line. Then decide on a timetable: month number 1, contact all training departments; month number 2, contact marketing departments, etc.

Put your best cost-effective foot forward. Managers should know the average length of programming, average costs of production, average cost of post-production, average duplication costs and average duplication run size. Managers should also have a firm idea of year-to-year aggregate costs and be ready with hard numbers in answer to hard economic questions.

Develop a marketing package. Develop promotional material. A natural, of course, is a five-to-ten-minute video promotional piece on the department: what you do, what you've produced, how you do it, with what results, etc. A print piece is also useful for wider distribution. And don't forget natural publicity outlets, like the company house organ, publications designed for special interest groups (e.g., safety directors) and local plant publications. Consider the media production department's physical surroundings as part of the marketing package. What does the department's environment say about how you do business, how professionally you do it and with what success? Here you might use photographs on the walls of the office or reception area to show off past successes and, of course, a display of awards won for media production excellence. The style of the marketing package says something about the department—it indicates the professionalism of the department's production techniques and the success of its programs.

9

Working with Executives, Clients and Experts

The success or failure of a media production department can greatly depend on the quality of interpersonal relationships between media production personnel on the one hand, and executives, clients and subject experts on the other. In many ways, media production operations are people-intensive. There are dozens of situations in which media staff interact with executives, clients and experts: during pre-production meetings, script reviews, storyboard sessions, video and film shoots, editing sessions, program reviews, budget reviews, problem-definition meetings, photography sessions, and more. It is important, therefore, that every member of the media staff be aware of the dynamics that can occur between themselves and others. It is essentially a matter of handling conflicts, conflicts that occur in various ways.

In order to analyze the kinds of conflicts that can occur and ways in which media personnel can handle these, I will describe a scene. This scene is the rehearsal of a management communications videotape session.

SCENE—THE REHEARSAL

10:00 am: The technical crew has been at work since 8:30. The video engineer has been adjusting the cameras for an hour. The director has made last minute adjustments to the lighting with one of the technicians. The producer is reviewing his pre-production notes. This morning's video recording will be used to help participants improve their on-camera performance and the program's structure and content. The subject of the program is energy conservation. The participants include the company's director of energy conservation, a corporate vice president in charge of energy supplies and materials services and the company president. Of the three, only the company president has been on television before.

10:05 am: A representative from the public affairs department walks in. He chats with the producer about the pre-production meeting of two weeks ago. He is nervous. He smokes two cigarettes in five minutes.

10:16 am: Nervous frowns turn to nervous smiles. The president and the others arrive. The rehearsal begins. The producer greets the president with a handshake:

''Warren.''

''How long are you going to keep me under the hot lights?''

''Not long. We should have you out of here in about 45 minutes. Warren, would you sit here at the far end of the table. You'll be talking into that camera for your introduction and close. Jim and Bruce—would you please sit on the long side here, and here.''

Bruce B., the energy conservation director, a man 6′3″ tall, hesitatingly asks, ''Should I be looking at the camera, or speaking to Warren and Jim?''

The producer replies, ''Your instincts are right. You should all be having a conversation with one another. The audience knows that this has been rehearsed and structured. They take that for granted. You know what you want to say. Warren, did you get that list of questions to ask Bruce and Jim, and the opening and close?''

''Yes, but I haven't had a chance to read them yet.''

''Why don't you take a moment now to look them over.''

The vice president asks the producer, ''What do I do about my very obvious, uh, uh, Southern accent?''

The vice president also has a stutter. The producer replies, ''Well, your accent is a part of you and you seem to have done well with it for a long time. Don't worry about it. Be yourself. That's the best thing.''

10:31 am: The executive cast finally settles down. Tape rolls. The director instructs the man on camera one to signal the president to begin his introduction. The president doesn't quite see it, despite the instructions given him earlier. He finally sees it on a third try.

The first run-through is dull and stiff. All three executives have misconceptions about how to act on television. They pose, they don't move their bodies, they look serious all the time. They do stick to the script's structure, though, which is positive. All the pre-production work has paid off in that respect.

10:55 am: The first rehearsal is over. The lights are shut off. The cast looks a bit perturbed. They don't know what to think of their performance. The director quickly rewinds the videocassette. He hands the cassette to the producer who puts it into a videoplayer. He pushes the play button and watches intently with the rest of the cast. He says nothing through the playback of the first few minutes. The president looks at the producer with a puzzled look.

"Well, what do you think?" the producer asks.

The president breaks the ice. "I think we need to loosen up."

A sigh is let out that is heard around the world. The two other executives willingly agree. The producer supports the conclusion. "Of course, the first time through is always the hardest. Everybody who has ever been on television goes through the same thing. The first time I saw myself on the tube, I thought, 'Is that me? Is that what I look like, sound like?' "

The producer observes more relaxation on the collection of faces. "Frankly, though, I will say that I don't think we need to make any changes in the content or structure. It plays out very well." The public affairs representative readily agrees. He suggests two minor structural changes in the body of the vice president's text. There is much agreement. The producer jumps in to initiate the second run-through before there are any more suggestions for minor changes.

11:13 am: The second rehearsal commences. It is much better, tighter. The cast is more relaxed. The vice president's stutter is far less pronounced. The president relates anecdotes to make the link between corporate energy conservation and energy conservation in the home. It is much more believable and human.

The second run-through ends at 11:30 am. The producer quickly plays back the second tape. They only watch the first five minutes. Everyone can see the improvement. Lots of smiling faces around the room.

11:37 am: Confirmation of the meeting place and time for the shoot tomorrow. The president and others leave. After a few closing words with the public affairs representative, the producer sits down for a debriefing with the technical crew. "We will probably shoot it in one take tomorrow," he says.

WHAT HAPPENED?

Identifying the "Teams"

Rehearsing executives for a management communications video program is much like a meeting between subject experts and media production personnel to discuss a forthcoming teleconference or a pre-production meeting between in-house corporate clients and still photographers to discuss a shoot for a planned product brochure. Each of these situations provides an opportunity to analyze some inherent situation conflicts between "management" and "functional" rank.

In the scene described above, there were various "players" present.[1] First is the producer. This individual is usually not of high managerial rank. In relation to the rest of the characters at the rehearsal, the producer stands above the technical crew (production engineer, director and the two camera persons) in managerial rank, but below the corporate executives.

On the other side is the public affairs department representative. This individual has more status than the producer primarily because he interacts more often with highly placed executives. Several notches above the public affairs representative is the director of energy conservation who, in turn, reports to the corporate vice president in charge of energy supplies and materials services. The president, of course, is highest in rank among all the performers in this scene.

Thus, there are essentially two teams in this drama. The producer and the technical crew are on one team; the president and the other executives are on the other, along with the public affairs representative.

Although the producer ostensibly belongs to the former team, he must also serve as the single member of a third team; that of ombudsman between the two teams. Therefore, the producer may play with the technical crew, but he may also have to play with the other team when appropriate, or alone when necessary.

Analyzing and Resolving Conflicts

From the outset, the scene is fraught with conflict. To begin with, the unstated but obvious conflict is (1) the differentiation between the managerial rank of the president's team and the producer's team, and (2) the "functional" ranking of the two teams. The producer's team may be low in rank, relatively speaking, in terms of managerial authority, but the president's team is completely outranked by the producer's team in terms of the skills required to perform successfully at the video rehearsal.

Thus, the president's team is put into the functional role of students—they don't know what to do in the rehearsal. The producer's team, on the other hand, is well versed on what goes on during a rehearsal; they have done it before with many other executives for similar kinds of programs. However, only the producer has the skills necessary to guide the technical crew and the president's team through the experience. Therefore, the producer must directly play the role of teacher.

Despite this inherent conflict, the motivation for both teams in this scene is ostensibly the same: to perform well at the rehearsal and to achieve a future performance that will be both believable and successful. However, there are variations on how each team achieves this objective. The producer's team must do all it can to ensure a technically successful rehearsal; i.e., properly focused lights, even audio levels, smooth camera work, properly timed cuts between cameras. The producer's task is to upgrade the performance of the president's team without overstepping the bounds of rank or personally insulting any of the president's team players. The president's team members, on the other hand, must maintain the semblance of managerial authority and skill (in this case, communication skills) associated with high rank, while at the same time maintaining continuity of program structure and content. Moreover, there is an added conflict in the president's team: they wish to know how to perform better in the unknown rehearsal situation, but cannot, in most cases, bring themselves to directly ask, "What should I do?"

It is debatable which team experiences the greater pressure. Perhaps the home team has a slight advantage. Not only does the producer's team have the psychological advantage of technical skills, but it also knows the physical territory. The president's team , on the other hand, must not only quickly acclimate to the new performance situation, but must also rapidly adjust to the unfamiliar physical environment.

The conflict between managerial and functional rank is even more pronounced during the rehearsal. There is an inherent conflict as a result of varying degrees of skill; also, the learning of the skill is very public. Even though the public is relatively small (the technical crew, the producer, the public affairs representative and the members of the president's team), each member of the president's team must impress the others by giving a good performance—and the successful performance must be done in the present, the here and now! A bad performance cannot be hidden. The president, too, must perform successfully, if he is to maintain his reputation.

Although his performance will not appear on videotape in front of a large audience, the producer's success is contingent on the successful performance of the president's team. The public affairs representative, however, has a couple of "play options." If the producer's actions result in a poor performance on the part of the president's team, he can always opt to say, "Perhaps we should have gone to a more professional organization to get this done," or "Video communications is too risky, let's try something else to communicate our message." If the performances are successful, he can opt to say, "I knew this would turn out well."

The rehearsal is an intriguing situation. Not only is it a means of improving one's future performance, it is also in itself a meaningful performance. As students of playwriting know, a good play is one in which there is a conflict and one in which one or more of the characters change in some way. In the rehearsal the essential conflict is between managerial and functional rank. And in more ways than one, if the producer performs well, the president's team members will leave the rehearsal changed; they will have become better performers, at least in front of the camera.

TECHNIQUES

The rehearsal for a video communications program is only one common situation in which executives, clients or subject experts come in contact with the media production department staff. Let us look at some other potential conflicts the media staff may encounter.

A subject expert or client walks into a producer's office and declares, "I want to produce a videotape. Can you help me?" If you probe the expert at this point, you'll find that he/she (especially if this is his/her first video production) is probably scared to death just walking into the office. Why? Video is a sophisticated electronic communications tool. Lack of knowledge of how video works or how it should be approached can be highly intimidating. The expert will probably try to maintain as much control of the situation (i.e., the production) as possible to keep from showing the fear he/she is probably experiencing. What should you do?

- At the beginning, let the client or expert do the talking. He/she may feel the need to convince the producer of his/her expertise. Be patient. The client or expert comes to the media production department with needs, not vice versa.

- Initially, avoid using too many technical terms. Converse in the client's language. Ask for definitions—people love to talk about what they know. The more he/she talks about the familiar, the more comfortable and confident the client will become. Ask questions about the expert's background. The more one knows about this person as a person, the more one will be able to deal with the expert on a personal level.

- Be patient. Convincing others that you know what you are doing takes time. No one likes to have totally new ideas thrust upon them at once and be expected to act on those concepts immediately. Let the client or expert assimilate your credentials slowly. You'll reach your objective faster.

- Be yourself. While it may appear that some people have become successful because they were the epitome of perfection, people prefer to deal with someone who is human, genuine and at ease with himself/herself.

- When dealing with clients, ignore *what* they are and concentrate on *who* they are. Clients want to know they are dealing with a professional, especially one who is not intimidated by rank.

- There is a need to reach a clear agreement about expectations concerning when and how things get done. If a client's expectations are too high—given the time, manpower and financial resources available—they must be lowered. Conversely, if they are too low—because of concern about the department's abilities or fear that the production process will be an agonizing one—then those expectations must also be changed before the process goes too far.

- Give clients what they need, not what they want. Sometimes clients ask for or expect things by expressing them as vague notions or stubborn demands. Sometimes a client will want something that is inappropriate; at times a statement is made in a video program, or article or speech draft, that upon later review is irrelevant or extraneous. Some clients may insist that their every "pearl of wisdom" remain, but common sense may dictate that the statement should come out. If the project is to be successful, an audience other than the client will have to receive it. Media staffs have an obligation to the client's audience to take out extraneous material.

- Clients, especially those uninformed about the media production process, are occasionally eager to show how much they know about producing. They feel they have to maintain command of the situation and begin the discussion by describing their ideas for script, set, graphics, metaphors, etc. While an immediate reaction could be to ask about the client's objectives and audiences, a useful riposte to this kind of client is to say "Well, before we start discussing style, let's talk about content; that is, what needs to be said, the order in which it has to be said, and who will say it." This can be an effectively disarming technique.

- Be selective as to when, and how involved, a client gets in the production process. A client does not have to be involved in every step. It is a rare client who will take as strong an interest in the production process as media staffs. However, because the large majority of clients are members of management, it is to their advantage that the time they spend in the production process be used effectively and efficiently. While the client should not become as totally immersed in the production as you, the client should still have opportunities to contribute to the program. However, if the suggested contribution is inappropriate, this should be made known, and reasons given. Again, what a client thinks is needed may not support his/her objectives, and these suggestions must be dealt with immediately.

- Keep the situation informal and unfettered by bureaucratic requirements. Good humor is a must. Minimize the amount of paperwork generated during the production process. Program proposals and budgets can be effective even when they are short. Schedules do not have to be complicated to be useful.

- Maintain a relationship with a client even after project completion. If the program was successful, the client will be back for more. Success breeds success.

Basically, to handle clients, one often uses proven salesmanship techniques. First, you have to sell yourself. But you really cannot do that until you know something about the customer; here, media clients. Once you have laid the foundation you can begin to sell your expertise, and thus reach the ultimate objective: satisfying the client's needs while meeting the audience's demands.

All in all, the best way to ensure that a media project is shepherded efficiently from conception to completion is to develop a client's respect for the media production department's talents and expertise. It is that simple. The more trust and confidence a client has in the department's skill and experience, the more likely the project will be successful, because the easier it will be to do what professionalism dictates should be done. Trust in the department's abilities will provide for greater flexibility in scheduling, making changes and explaining why something didn't get done when planned. Creating this trust is the difference between a media production department that is merely competent and one that is successful.

FOOTNOTE

1. The dramaturgical analogies used in this analysis are borrowed from Erving Goffman's *The Presentation of Self in Everyday Life* (New York: Doubleday Anchor, 1959).

10

Using External Resources

Striking an appropriate balance between an organization's use of in-house and external resources is a fundamental key to the effectiveness of a corporate media production operation. Media managers should use external resources when appropriate, and coordinate these resources with in-house capabilities. In my experience, external resources are indispensable, no matter how advanced the department is. Often, the more work an in-house operation does, the more it tends to use external resources.

OUTSIDE RESOURCES TO CONSIDER

The following is a list of the types of external resources a media department may consider using:

Consultants

organizational communication systems consultants
graphic design consultants
film production consultants
video production consultants
audiovisual facilities design consultants
talent agencies
photographic facilities design consultants
video production facilities design consultants
teleconference facilities design consultants
videodisc programming consultants

Production facilities

typography services
print duplication production
photography retouching and color correction
film editing facilities (16mm and 35mm film)
film laboratories
film duplication (super 8mm, 16mm and 35mm)
film (audio track) mixing
film animation stand production
video production
video post-production
videotape duplication
audio production
audiotape post-production
audiotape duplication
audiotape mixing
equipment rental (for camera, lights, etc.)
slide duplication
multi-image production
slide-to-film transfers
slide-to-tape transfers
film-to-tape transfers
tape-to-film transfers
16mm film to super 8mm film duplication
multi-image to videotape transfers
multi-image to film transfers

Production houses

A production house differs from a production facility in that it provides the creative staff, along with—but not always—the production facility.

graphic production
film production
slide production
video production
multi-image production
audio production
photography production
videotape production

Freelancers

art directors
graphic designers
mechanical artists
photographers
cinematographers
directors of photography
lighting directors
producers
directors
audio recordists
voice-over professional talent
maintenance engineers/technicians
print copywriters
film scriptwriters
video scriptwriters
multi-image production scriptwriters
videodisc programming scriptwriters
on-camera professional talent
camera persons (video)

models
film animation/photoanimation designers
film editors
video editors
video engineers
audio engineers
production assistants
technical directors
researchers
administrative assistants
music editors
stage managers
audiovisual technicians
grips and gaffers
makeup artists
set designers
costume designers

Canned materials

music and sound effects
canned films
canned video programming
film stock shots
video stock shots
slide libraries
canned multi-image programming
picture (graphic) libraries

Supplies

paper
pens, brushes, and pentels
photographic paper
audiotape
videotape
positive and negative 16mm and 35mm film
mag track (for film production)
bulbs
equipment parts
photographic chemicals

Equipment (hardware) suppliers

graphic storage
graphic design tables
still photography cameras
photographic processing
slide production stands
slide duplication
cinemagraphic equipment
video cameras
video editing
videotape recorders
film projectors
audio recording and mixing equipment
lighting equipment
videotape duplication
distribution amplifiers
film recording
audio mixing
teleconferencing

CRITERIA FOR SELECTING EXTERNAL RESOURCES

The organization must have some criteria for selecting external resources, and these will vary depending on the specific resources involved. Below are some suggested criteria for each external resource mentioned.

Consultants

What criteria can the organization use for determining which consultant is the best for the job? Here are a few:

How many years of experience does the consultant have?
Does he/she have a successful track record?
Has the consultant worked for one kind of organization, or for organizations in different industries?
What is the consultant's reputation in the industry and with former clients?
What kind of organization does the consultant have?
What is the consultant's operating philosophy?
Is he/she open to new ideas, or willing to apply new technologies?
Will the consultant work well with others in the organization?
Does the consultant have an inquiring mind?
Is the consultant willing to listen to others, and is he/she flexible?
Is the consultant familiar with the work of other leading consultants in the field?
What kind of professional organizations does the consultant belong to?
Is the consultant available when needed?
Is the consultant's fee structure competitive?
Does the consultant communicate ideas clearly and succinctly? Both verbally and in writing?
Is the consultant willing to prepare a written proposal?

Production Facilities

With respect to production facilities, the following questions are crucial:

What kind of equipment does the facility have, and is it well-maintained?
Is the equipment up-to-date or state-of-the-art?
Is the equipment serviced by a professional staff?
What is the facility's reputation with existing customers?
How long have they been in business?
Does the facility have a reputation for a high level of service and cooperation?
What is the facility's overall level of quality?
What is the facility's price structure? Is it competitive?

Who are its clients?
Does the pricing schedule match the level of quality and service?

Production Houses

When choosing production houses, the following criteria might be applied:

How well acquainted is the production house with quality production facilities in the area?
How long has the production house been in business?
Does it specialize in one kind of application (such as marketing, public relations, etc.)?
What is its reputation with clients?
What kinds of organizations has it worked for in the past?
What is its style of production?
Does the production house have a reputation for producing programming that is on time, within budget and meets objectives?
Does it have a systems approach to media programming, or is it more interested in producing programming that is merely aesthetically pleasing?
How well will production house personnel work with in-house clients, subject experts or executives?
Is their price schedule competitive?
How many examples of previous work is the production house representative willing to show?
What kind of staff does the production house have?
What is the depth of experience of the production house principals and staff?
Does the production house have a successful track record?
What is the production house's reputation in the industry?
Has the production house been written up in trade publications?
What is the production house's operating philosophy?
Is the production house open to new ideas, or does it use old formulas?
Does the staff have a businesslike approach as well as a creative style?
What standards of quality does the production house adhere to?
Is the production house staff willing to listen to others?
Is the production house familiar with the work of leading houses in the area, and with the work of leading organizational practitioners in the field?
Is the production house staff flexible?
Is the production house available when needed?
Does the production house principal communicate ideas clearly and succinctly?

Is the production house willing to present the organization with production ideas, budgets and schedules in writing?

Freelancers

Many of the same questions posed to consultants, production facilities and production houses could be asked of freelancers. In addition, for supplies and equipment, price, warranties and service are important criteria for selection. For example, if the external resource is a dealer in photographic or video equipment, the organization should question the dealer's ability and reputation with respect to delivery date commitments. The organization should also be concerned about the dealer's willingness and reputation to fulfill warranty commitments.

BIDDING AND PURCHASING

Pricing, of course, is an important consideration. Other things being equal, organizations should use a bid procedure when selecting an outside vendor, especially when equipment or supplies are involved. However, there will be times when the lowest-priced vendor may not necessarily be the best external resource for equipment and supplies. The organization has to examine the vendor's reputation, delivery record and service reputation. All these criteria together will determine whether the vendor should be given a contract. The organization's purchasing department should get involved, particularly when the external resource involves capital equipment and supplies. In these instances in particular an organization's purchasing function may have more clout in obtaining the right price and delivery schedule than the media department will.

Organizational purchasing functions can also get involved in hiring consulting, production facility or production house services. However, in these instances the purchasing function may serve more in an administrative capacity. The role of the purchasing function in these cases will differ from one organization to another. It should be underlined, however, that the purchasing function can be a useful ally regardless of the service being purchased, either before, during or after the service has been provided. In many cases, the purchasing function can serve the media production department well by helping to draw up purchasing contracts that protect the organization, while at the same time providing the external resources needed.

THE PRESENTATION

A marketing presentation by the external resource being considered may be a clue to the resource's desirability. An external resource can ''pitch'' itself through a print piece, a phone call, an on-site presentation, a visit by a representative or a combination of the above.

Quality

I have received presentations from hundreds of consultants, production facilities, production houses, freelancers and hardware suppliers. In general, the good suppliers differentiate themselves in several ways.

The first is quality: an in-house organizational client once told me that the first level of success is to perform a job effectively; does the product do what it is supposed to do? Now, if that effective job is also done efficiently (i.e., on time, and within budget) then you are really exemplary. Thus, the first point to consider is the quality of the external resources' wares. How do you judge quality? Here are some criteria:

Is the level of quality consistent, or does it vary from one job to another?
Does the effectiveness of the media presentation reach you on an emotional level?
Do you feel that this vendor is interested in trying just a little harder than the next guy?
Would you be proud to show this vendor's wares to others?
Do you feel the vendor is professional?

Of course, the question of what is quality and what is not is sometimes a matter of taste. There are, however, certain technical ways of judging quality, such as:

Are the colors in the print piece true, and in register?
Is the layout busy or clear?
Is the presentation clear and understandable?
Is the photography fuzzy or sharp?
Is the quality of the film grainy or sharp?
Are the graphics telegraphic or do you have to search the piece to begin to understand what you are looking at?
Is the quality of the audio track understandable, does it make creative use of music, sound effects, voice-overs?
Is the quality of the video picture sharp, and well lit? Is the camerawork creative or dull? Smooth or jerky?
Is the quality of the editing jarring or smooth, and in keeping with the content of the program?

Style

In addition to the technical quality of the vendor's product, the style of the presentation is also a criterion for selection. In my experience, the most successful presentations were those that were simple, well organized and short. The vendor representative did not try to overwhelm me with credentials, but gave just enough information to whet my appetite for more. The best presentations by production houses, consultants and freelancers have been those where the presentation of product was both broad and short. In other words, I was presented with a variety of products, each of which was a small part of the total product (as opposed to being asked to sit through a 30-to-40-minute film on a subject that had nothing to do with my clients). The question of variety is especially important. If an outside vendor shows up with only one or two examples of work, this vendor should be suspect.

Careful Analysis

The better outside resources, consultants in particular, are those who will not be stampeded into giving a price at the presentation. They may, however, be willing to talk about rates or ball-park figures. The better resources will always ask to have more details about a particular project before committing themselves to a price. In my view, this attitude says several things. First, the vendor is not willing to give his product away; he is not desperate. That reflects a certain level of confidence. Second, the vendor is willing to spend some time working out the details of a particular project. In the long run, this usually means a higher level of quality and service.

Price

The lowest-priced vendor is not necessarily synonymous with poor service, performance or product. The reverse is also true: high price does not guarantee high quality. The more professional external resources will be willing to submit fairly detailed price breakdowns, as well as schedules of performance.

WHEN TO USE EXTERNAL RESOURCES

External resources should and will be used most of the time, no matter what the size, scope or level of media production activity. Even if the media operation consists of three people—one graphic artist, one photographer and one video producer/director—all three will require supplies in order to perform their respective functions. At the other extreme, a larger media operation—will very likely be in constant need of external resources, such as consultants, production facilities, production houses, freelancers, supplies, canned materials and hardware vendors.

The trick to using outside resources effectively is matching the right kind of external resource at the right time for the right project at the right price, together with internal resources (while at the same time charging the organization a reasonable price for services rendered by the internal production resource). The combination of the two resources—internal and external—should at one and the same time provide the organization with cost-effective media production services, while keeping the in-house operation competitive, i.e., cost-effective.

To a certain degree, the decision to use external resources may have to be made on a project-by-project basis. External resources will come into play at different times for different reasons (such as when the in-house operation expands and more media execution is accomplished in-house).

Consultants

Consultants can be used at various times in the development of a media production operation, for example, if an organization is about to change its headquarters location, reorganize, change its product line, change its pricing structure or revamp its internal or external communications activities, consultants can be called in to help develop ways of effecting this change. Here, media production consultants can be very useful, especially if they have a broad range of production experience (beware of production consultants who have never gotten their hands dirty, technically speaking). Facilities design consultants should be called in if hardware equipment and/or facilities are going to be considered, installed, revamped or changed from one system to another (such as from videotape to videodisc playback). Experienced consultants should always be called in if a new technology is under consideration, even if in-house personnel are knowledgeable about it and are well prepared to use it. The consultant in this instance serves as a sounding board and provides a knowledge of other organizations, experiences against which the organization can judge its own potential use of the technology.

Production Facilities

There are several instances when a media department should use outside production facilities. The most obvious is when the organization has no media production facilities of its own. Outside production facilities should also be used if the volume of media production is not high enough to warrant the installation of in-house production equipment. Another reason for using outside production facilities is when the external facility has state-of-the-art equipment that the in-house media production organization cannot purchase because its frequency of use is not high enough. In some cases, the volume of in-house work may be so high that overflow must be handled at an outside facility. Finally, scheduling conflicts and deadlines sometimes make it necessary for work to be performed at an outside facility.

Production Houses

Many of the reasons for using production facilities apply equally to using production houses. If an organization has no in-house media production talent, then an outside production house must be considered. Moreover, if the volume of media production work is not constant or high enough for organizational management to consider hiring creative or technical talent, then outside media production houses should again be considered.

Even when an organization has in-house creative and technical staff, outside production houses may be more up-to-date, since they are in contact with many organizations. This contact requires the production house to be not only competitive, but also up-to-date in order to remain so. Production houses can be useful when the volume of work is so high that in-house staff cannot handle the overflow. Media production schedule conflicts may force the use of outside production houses. An additional reason for using outside production houses is when the nature of a project is such that the in-house staff is not positioned to handle it. For example, the in-house media production group may be organized to effectively and efficiently handle media production that is daily, weekly or even monthly, but some projects may be long-range and require extensive traveling. While an in-house representative may be designated to oversee the project, an outside production house may prove more cost-effective in the long run.

Freelancers

Freelancers should be used almost constantly, no matter how large or small the media production operation. The basic reasons for using production facilities and production houses apply equally to freelancers:

- When the organization has no in-house creative or technical media production talent.
- When the volume of work is not sufficiently high to warrant the

hiring of in-house creative or technical media production talent.
- When the in-house staff is not as up-to-date or professional.
- When the volume of work is so high the in-house staff cannot handle that work.
- When in-house scheduling conflicts necessitate the use of freelancers.

There are several other times when it proves more cost-effective to use freelancers. In certain instances freelancers are more cost-effective than in-house staff. Here, an important criterion is volume. If, for example, the organization only occasionally originates graphics, photography, films or video programs, freelancers will prove more cost-effective. Similarly, there are certain production functions that do not occur on a frequent basis. For example, unless an organization is producing video programs by the hundreds every year, it does not pay to have camera persons on staff. The same holds true for film editors, lighting designers and makeup artists.

Freelance scriptwriters present an interesting situation. I have found that it is not only more cost-effective to use freelance writers, but also more communications-effective. Most media production departments do not have scriptwriters on staff; this function usually falls to the producer/director, who is often handling several media projects as well. Therefore, making use of a freelance scriptwriter frees the in-house producer/director, allowing him/her to be more productive. Moreover, other things being equal, if the producer/director were to figure out the cost of his/her time to write a script, freelance scriptwriters might be less expensive. Finally, no one scriptwriter can effectively write for all kinds of communications applications. Some writers are more adept at writing marketing programs than training programs, and vice versa. With the advent of the videodisc the difference in scriptwriting expertise among various applications is even more pronounced.

Canned Materials

Canned materials present an obvious opportunity for the organization to save money. Implicit in the term "canned" is the idea that the material has already been produced; the organization does not have to originate the material. There are a myriad of companies that have produced videotapes and films on a variety of subjects. The availability of canned programming does not preclude an organization from originating its own material, of course. It may even modify canned programming, provided the organization receives written permission from the program originator.

Canned media elements, such as music and stock shots, can be found through media libraries and contribute to original programming. When using canned materials, it is important to obey the copyright laws. The originator of the material presumes it will be used in accordance with copyright law.

Hardware and Supplies

Once an organization decides to initiate, install and develop a media production department, it needs to develop relationships with outside hardware and supplies vendors. This is true for small, medium and large operations.

DEALING WITH EXTERNAL RESOURCES

These are some additional pointers to keep in mind when dealing with external resources.

Don't wait until you need a vendor to find one. Check out the availability of vendors in your area. Know who to call on in an emergency. Get to know the best vendors in your area. Keep a file and review it from time to time. Take nothing for granted: if a vendor looks good, check out his credentials. Ask other organizations what kind of service the vendor provides. Keep the file up-to-date. You may be satisfied with your current vendor, but continue to meet new ones to keep the pool fresh.

Don't put all outside needs in one vendor basket. No single vendor can satisfy all outside needs. Segment your needs. For example, it may be less expensive to use an outside production facility, hire a freelance writer for the script, and another freelancer to produce and direct it. Using different vendors for hardware and software might help bring costs down.

Don't expect miracles. Most outside suppliers who survive the marketplace survive because they are good and they have something to offer. However, they are just as susceptible to having problems and making mistakes as those who are part of the in-house operation. Moreover, working with an outside vendor may mean working just as hard to get what is needed as working with in-house resources.

Demand quality. People usually only give what they perceive others will demand of them. Ergo, outside vendors will usually only provide that level of quality they think you will expect. Demand quality and good outside vendors will respond. Sometimes, challenging vendors to perform at a higher level than expected proves fruitful not only for the organization but for the outside vendor as well.

Let good vendors know you appreciate their work. Outside vendors are not the enemy. If they do a good job, let them know it (and if they don't, let them know that too, even when it's an exception). Everyone needs feedback, both positive and negative. All that is needed is a phone call or a short note. Either way, effective relationships with outside vendors (just like with in-house clients) need to be cultivated. Sharing feedback is one way.

Confirm agreements in writing. No matter how long-standing a relationship with an outside vendor, confirm agreements in writing. The legal department does not need to get into the act every time you call up a supplier for a box of videocassettes. But it does mean that a confirmation of the work to be performed should be in writing in some form. A good outside vendor will do this as a matter of course. Many organizations require an outside vendor to supply a written estimate as a check against actual charges. Something in writing up front helps answer a lot of questions later.

Pay vendors on time. A paid vendor is a happy vendor. Just imagine if the organization didn't pay in-house personnel on time!

Trust your vendor. Once you've established an ongoing relationship with an outside supplier, trust him/her. If you're constantly calling to find out if the script is finished or if the shipment of tapes has left the warehouse, you will only create an unnecessarily tense situation. If you don't feel the vendor will give you the kind of service you want without constant double-checking, find a vendor who will.

Learn your vendor's business. Ask vendors questions about how they do what they do for you. Understand their problems while they try to solve yours. Become fluent with the terms of each vendor's profession. In-house personnel often turn to outside suppliers for help with a problem, and therefore must be able to communicate effectively with them.

SOURCES OF EXTERNAL RESOURCES

There are a variety of ways to find the external resource appropriate for the organization. Some of these are discussed below:

Vendors. Outside suppliers are in constant contact with others in the business, not necessarily just in their line of service. Thus, hardware vendors may know the name of a good maintenance technician, a scriptwriter may know the name of a good technical director, a graphic artist may know the name of a good director, and so on.

People in the business. Organization personnel can also be sources of potential external resources. One department member may be in contact with certain external resources that may be appropriate for use by another. Other organizations, likewise, can be a valuable source of information on the location, availability and desirability of external resources. Headhunters, professional organizations and institutions of learning can also provide leads.

Publications and trade shows. External resources very often advertise, are written up or publish articles in trade journals and publications. Some may also advertise their services in consumer publications. Both sources can be used as a means of finding the appropriate external resource. Trade shows are prime sources for learning about the suppliers who exhibit there and are happy to show their products. *Folio* (New Canaan, CT) aimed at magazine and book publishing, is a good source of printers, for example, and *Video Expo* (Knowledge Industry Publications, Inc., White Plains, NY) is one way to find out about video equipment and applications. Others include the Visual Communications Congress, the NAB (National Association of Broadcasters) and ICIA (International Communications Industries Association). See Appendices II and III for addresses.

11

Contracts, Copyrights and Permissions

In the previous chapter, the media production manager's use of external resources was discussed. The use of an outside resource immediately raises important legal issues. For example:

- Once a consultant's work is completed, who owns the work—the consultant or the corporation?
- When a program is finished, does the production house own the materials that went into the creation of the presentation?
- Who is responsible for providing on-location insurance in the event of accidents?
- Who is responsible for the loss of materials such as master videotapes?
- What is the media producer's legal responsibility with regard to the use of canned music and sound effects, slide and stock photography?
- What are the media producer's responsibilities toward professional actors (in photographs, slide shows, videos, films et al.)?

CONTRACTS

The moment a media producer hires someone from outside the organization—a consultant, a production facility, a free-lancer, an actor or narrator—or licenses existing materials for a production such as canned music or stock photography, a contract is created between the producer and the outside resource.

In some cases (e.g., professional talent) pre-negotiated contract conditions and minimum fees have been determined by collective bargaining. In other instances (such as when the organization hires a consultant, free-lancer or production facility) each situation requires individual negotiation.

In either case, the contract "between the parties" should be determined in advance and specified *in writing!*

If a consultant, free-lance producer or director, writer or photographer is hired on a project basis and the terms and conditions of the project are not set forth, or are written down but articulated poorly, there can be problems either during or after the project is finished. The organization's legal and purchasing departments can be useful allies in drawing up contracts with outside resources.

Perhaps you are thinking: "All I want to do is hire a production assistant for the day. Does this mean I have to draw up a 20-page document discussing all conditions and contingencies?" Of course, the answer is no. But there should always be some confirmation of the terms and conditions of the work to be performed by the outside resource in writing. With respect to a production assistant a simple letter stating:

> This is to confirm you will provide XYZ Corporation with production assistant services on July 14, 1989 for a fee of $100. Start time is 8 a.m. We should complete the production by 5 p.m. You will provide your own transportation to the location.

should suffice.

If, however, the organization hires an outside production company to produce a 20-minute videotape for a budget of $150,000, then a far more extensive "confirmation of services to be performed" is required.

The contract, for example, should specify the purpose and content of the program, the production schedule (daily and weekly), the production facilities and services to be provided, production principals and their credentials, errors and omissions, insurance and a fee payment schedule. Certain organizations require outside production companies to agree to provide Workers' Compensation for statutory obligations, liability insurance and automobile insurance. In certain instances, the hiring organization will be named as additional insured to the production company.

Companies may state they can terminate the contract at any time prior to the completion of the work; further, that the work will be the exclusive property of the company. Other standard (so-called "boiler plate") clauses might include the following:

> All data, reports or other written records of the work, and all information pertaining to the work, shall be considered proprietary to the company and held in confidence. The production house will not publish or otherwise disclose to others such data, reports, written records of information without the express prior written consent of the company.

Or

> The work may be required for use in a contest adversarial proceeding. The production company will not comment upon, discuss or otherwise communicate to anyone, any matter concerning the work including the nature and scope of the work and the fact that the production company is rendering services to the company without the express prior written consent of the company.

The contract might also include clauses dealing with Equal Employment Opportunities, Payment of Taxes (when applicable), Other Application Laws and Regulations, Other Labor Laws or Collusion or Fraud.

A well-written contract will also specify the amount of fees to be paid and the schedule of payment. In some instances, the organization and the outside resource may agree to a "fixed price" contract. This means the price of services to be rendered is fixed, regardless of circumstances.

It could also be agreed that the outside resource will perform a service for "cost, plus." This means that even though a price may have been set, circumstances may dictate additional fees. For example, if a photography project lasts three days instead of two, then most certainly this adds to the cost of the shoot. If a client makes changes in a slide show after it has been designed and executed, this adds to the cost. If changes are made to a videotape after editing has been completed, this, too, will increase the cost of the project.

It should, therefore, be agreed up front when changes and what kinds of changes will result in additional fees. With respect to a videotape, an organization should expect additional fees if changes are made:

- after the script has been reviewed and approved
- after pre-production planning has been completed
- after a scout has been completed

- after casting has been completed
- after a voice-over has been recorded and approved
- after music has been selected, reviewed and approved
- after a rough cut has been reviewed and approved
- after the online edit has been completed, reviewed and approved
- after the music/voice-over mix has been completed, reviewed and approved.

Needless to say, as with the planning of a media production operation, good planning of a media project, including specifying the terms and conditions of a contract with outside resources, will preclude problems that could arise during the course of a project.

COPYRIGHTS AND PERMISSIONS*

Under the old copyright law of 1909, a copyright was considered a single bundle of exclusive rights. It regarded copyright as an indivisible entity, which meant that a transfer of less than the entire package of rights to a work was merely a license allowing a holder to use the work in a specific way but did not permit him/her to exercise any right of ownership. Unless there was a specific agreement signed to the contrary, the one who hired an author, for example, was deemed the copyright owner.

The "new" law, which went into effect on January 1, 1978, contains the first explicit statutory recognition of the principle of divisibility of copyright ownership in U.S. law. Its impact is that any of the exclusive rights that make up a copyright can now be transferred and separately owned. This divisibility applies whether or not the transfer is limited in time or place of effect.

Under the old law, in the absence of an agreement to the contrary, the person or entity who hired an author owned the copyright; under the new law, the shoe is generally on the other foot.

The law also provides a term of copyright which lasts the author's life plus fifty years. The law specifically confirms a principle known as "fair use," which sets limitations on the exclusive rights of copyright owners. The law created a Copyright Royalty Tribunal whose job is to determine what rates are reasonable in certain categories.

Copyright extends to all kinds of literary and artistic creations, including music, novels, poems, drama, "audiovisual" works and graphic arts, in

---

* Much of the information in this section on copyrights and permissions is taken, with permission, from the *Stock Photography Handbook*, American Society of Magazine Photographers, New York, 1984, pgs. 138-140.

a virtually endless list. Each different type of work may be used in many different ways. For example, a drama may be duplicated (printed copies run off), performed (on the stage), or recorded (on film or videotape).

Therefore, it is frequently said that copyright is a "bundle of rights," protecting these many different types of uses, singly or in combination. That bundle is infinitely divisible: each right may be divided and subdivided, to allow for the appropriate economic gain from each different type of use of the property. Put another way, copyright enables the author to control and be compensated for each protected use of his property through appropriate licensing arrangements.

Photography

The use of stock photographs is protected by copyright law. Once a photographer shoots a picture, he owns it. He can give all or part of it away only by a signed agreement. In most transactions he may license a right rather than give away ownership.

Under fair use, a party may utilize a photograph which is owned by a photographer for such purposes as criticism, news reporting, teaching, scholarship or research. This is known as fair use and no permission or other authorization is necessary. However, in determining whether the use is considered fair, the following factors are taken into consideration:

1. The purpose and character of the use, including whether such use is for commercial or nonprofit educational purposes.
2. The nature of the copyrighted work.
3. The amount and substantiality of the portion used in relation to the copyrighted work as a whole.
4. The effect of the use upon the potential market value of the copyrighted work.

Thus, if a photograph is being used without permission, the photographer would first establish whether or not the use might be considered fair use. The above criteria should then be reviewed.

Limited reproduction by libraries and archives is permitted provided the reproduction is not made for commercial purposes.

The law provides that noncommercial transmissions by public broadcasters of copyrighted published graphic works (such as photographs) are subject to a compulsory license, and that if the copyright owners cannot agree on the terms of the license, the Copyright Royalty Tribunal can fix the royalties.

What happens when a corporation uses a copyrighted photograph without permission and the use falls neither under the fair use provision nor the compulsory license section? The law is quite specific on this. Anyone who violates these exclusive rights is deemed an infringer. The photographer can sue.

Music*

Music plays a significant role in many corporate media presentations. In fact, the use of music in a wide variety of communications applications is probably more extensive than usually realized. Whatever the application, most music—whether an original composition for the occasion, popular music used in a corporate context or stock music—is copyrighted.

Copyright, of course, protects musical works. Certain uses of music are more important commercially than others. Among the more significant rights in music are the rights to print sheet music, to make phonograph records and audiotapes, to "synchronize" or record the music in timed relation with visual images on a sound track of a film or videotape and, most importantly, the right to perform the music publicly.

Some musical works are in the public domain, for example, many traditional songs, and the classical works of Bach, Beethoven and Brahms. But while the song may be in the public domain, a particular recording of it may not be. And never take it for granted that a work *is* in the public domain—for example, "Happy Birthday" is a copyrighted musical work. If there is any doubt as to the copyright nature of the musical work, it should be investigated to avoid problems later.

In most instances, a corporate internal or external communications presentation will use stock music. However, in some instances, a company might use so-called popular music, either in its original form or with new lyrics. In either event, the music is copyrighted and the corporation has a legal responsibility for seeking out the copyright owner and negotiating a fee for the use of the material.

Public Performance of Music

When a company uses a piece of so-called popular music for a "public" internal or external communications project, e.g., sales meeting or seminar, that is copyrighted, the organization must negotiate a license for the performance of the underlying composition either directly with the music publisher or the composer (whoever holds the copyright or the rights of

* Much of this section on music is based on Bernard Korman's and I. Fred Koenigsberg's, "Performing Rights in Music and Performing Rights Societies," *Journal of the Copyright Society of the USA* 33, no. 4 (July 1986): 332-367. Copyright ©1986 by Bernard Korman and I. Fred Koenigsberg.

the distribution to the copyright).

Popular music is music that is generally performed in the public arena: at concerts, nightclubs, bars, et al. This is to be differentiated from the use of music in commercials, television programs or corporate presentations. This relates to so-called "synchronization rights."

If a company uses a piece of popular music in a nondramatic, public performance, either in its original form or with new lyrics or in a newly arranged form, the company must negotiate a license either directly with the copyright holder or through a performing rights society, such as the American Society of Composers, Authors and Publishers (ASCAP) or BMI. Figures 11.1 and 11.2 show sample ASCAP licensing agreements.

Licensing Users and Users' Liability for Infringement

Under the 1976 Copyright Act, virtually every user who publicly performs music must obtain a license from the copyright owner, or be liable for infringement.

For licensing purposes, the American Society of Composers, Authors and Publishers (ASCAP) divides users into two categories—broadcasters and "general" users. Virtually all broadcasters perform music, and so require licenses. ASCAP licenses broadcasters from its main office in New York.

Nonpublic Performance of Music: Synchronization Rights

Some corporate media presentations involve popular music that is used in a nonpublic context, such as in a marketing or sales or public relations presentation on video, film, or multi-image. In this case, the corporation will have to negotiate again either directly with the holder of the copyright or with an agency specifically set up to negotiate license fees with copyright holders.

Several agencies act as the licensing agent between the copyright holders and producers. The biggest is the Harry Fox Agency (New York City).

The Harry Fox Agency (HFA) was established in 1927 by the National Music Publisher's Association to provide an information source, clearinghouse and monitoring service for licensing musical copyrights. The agency represents more than 6000 American music publishers and licenses a large percentage of the uses of music in the United States on records, tapes, CDs, imported phonorecords, in films, on TV and in commercials.

HFA provides the following services in the United States on behalf of its publisher principals:

Figure 11.1: ASCAP Sample License Agreement for a Nondramatic Live Performance

Agreement between AMERICAN SOCIETY OF COMPOSERS, AUTHORS AND PUBLISHERS ("SOCIETY"), located at

and

("LICENSEE"), located at as follows:

1. **Grant and Term of License**

(a) SOCIETY grants and LICENSEE accepts for a term of one year, commencing , and continuing thereafter for additional terms of one year each unless terminated by either party as hereinafter provided, a license to perform publicly at each of the locations specified in Schedule "A", annexed hereto and made a part hereof, as said schedule may be amended as hereinafter provided ("the premises"), and not elsewhere, non-dramatic renditions of the separate musical compositions now or hereafter during the term hereof in the repertory of SOCIETY, and of which SOCIETY shall have the right to license such performing rights.

(b) LICENSEE agrees to give SOCIETY notice in advance of any additional premises owned or operated by LICENSEE where music is to be performed during the term hereof, and Schedule "A" shall thereafter be deemed amended to include such additional premises. Such notice shall include all information as to "LICENSEE'S Operating Policy" (as hereinafter defined) required for each of the premises by this agreement.

(c) Either party may, on or before thirty days prior to the end of the initial term or any renewal term, give notice of termination to the other. If such notice is given the agreement shall terminate on the last day of such initial or renewal term.

2. **Limitations on License**

(a) This license is not assignable or transferable by operation of law or otherwise, and is limited to the LICENSEE and to the premises.

(b) This license does not authorize the broadcasting, telecasting or transmission by wire or otherwise, of renditions of musical compositions in SOCIETY'S repertory to persons outside of the premises.

(c) This license is limited to non-dramatic performances, and does not authorize any dramatic performances. For purposes of this agreement, a dramatic performance shall include, but not be limited to, the following:

(i) performance of a "dramatico-musical work" (as hereinafter defined) in its entirety;

(ii) performance of one or more musical compositions from a "dramatico-musical work" (as hereinafter defined) accompanied by dialogue, pantomime, dance, stage action, or visual representation of the work from which the music is taken;

(iii) performance of one or more musical compositions as part of a story or plot, whether accompanied or unaccompanied by dialogue, pantomime, dance, stage action, or visual representation;

(iv) performance of a concert version of a "dramatico-musical work" (as hereinafter defined).

The term "dramatico-musical work" as used in this agreement, shall include, but not be limited to, a musical comedy, oratorio, choral work, opera, play with music, revue, or ballet.

3. **License Fee**

(a) In consideration of the license granted herein, LICENSEE agrees to pay SOCIETY the applicable license fee set forth in the rate schedule annexed hereto and made part hereof, based on "LICENSEE'S Operating Policy" (as hereinafter defined) for each of the premises, payable quarterly in advance on January 1, April 1, July 1 and October 1 of each year. The term "LICENSEE'S Operating Policy", as used in this agreement, shall be deemed to mean all of the factors which determine the license fee applicable to each of the premises under said rate schedule.

(b) LICENSEE warrants that the Statement of LICENSEE'S Operating Policy annexed hereto for each of the premises is true and correct.

(c) Said license fee totals Dollars ($) annually, based on the facts set forth in said Statement(s) of LICENSEE'S Operating Policy.

4. **Changes in Licensee's Operating Policy**

(a) LICENSEE agrees to give SOCIETY thirty days prior notice of any change in LICENSEE'S Operating Policy for any of the premises. For purposes of this agreement, a change in LICENSEE'S Operating Policy shall be one in effect for no less than thirty days.

(b) Upon any such change in LICENSEE'S Operating Policy resulting in an increase in the license fee, based on the annexed rate schedule, LICENSEE shall pay said increased license fee, effective as of the initial date of such change, whether or not notice of such change has been given pursuant to paragraph 4(a) of this agreement.

(c) Upon any such change in LICENSEE'S Operating Policy resulting in a reduction of the license fee, based on the annexed rate schedule, LICENSEE shall be entitled to such reduction, effective as of the initial date of such change, and to a *pro rata* credit for any unearned license fees paid in advance, provided LICENSEE has given SOCIETY thirty days prior notice of such change. If LICENSEE fails to give such prior notice, any such reduction and credit shall be effective thirty days after LICENSEE gives notice of such change.

(d) In the event of any such change in LICENSEE'S Operating Policy, LICENSEE shall furnish a current Statement of LICENSEE'S Operating Policy and shall certify that it is true and correct.

(e) If LICENSEE discontinues the performance of music at all of the premises, LICENSEE or SOCIETY may terminate this agreement upon thirty days prior notice, the termination to be effective at the end of such thirty day period. In the event of such termination, SOCIETY shall refund to LICENSEE a *pro rata* share of any unearned license fees paid in advance. For purposes of this agreement, a discontinuance of music shall be one in effect for no less than thirty days.

5. **Breach or Default**

Upon any breach or default by LICENSEE of any term or condition herein contained, SOCIETY may terminate this license by giving LICENSEE thirty days notice to cure such breach or default, and in the event that such breach or default has not been cured within said thirty days, this license shall terminate on the expiration of such thirty-day period without further notice from SOCIETY. In the event of such termination, SOCIETY shall refund to LICENSEE any unearned license fees paid in advance.

6. **Notices**

All notices required or permitted hereunder shall be given in writing by certified United States mail sent to either party at the address stated above. Each party agrees to inform the other of any change of address.

IN WITNESS WHEREOF, this agreement has been duly executed by SOCIETY and LICENSEE this day of , 19 .

AMERICAN SOCIETY OF COMPOSERS,
AUTHORS AND PUBLISHERS

By ________________
DISTRICT MANAGER

____
LICENSEE

By ________________

____
TITLE

Source: American Society of Composers, Authors and Publishers. Used with permission.

Figure 11.2: Sample ASCAP License for a Live Corporate Event

RATE SCHEDULE FOR MECHANICAL MUSIC - SEMINARS

ATTENDANCE	DAILY LICENSE FEE
Up to 100 persons	$15
101 to 250 persons	25
251 to 500 persons	35
Over 500 persons	50

The above rate schedule covers the use of mechanical music in conjunction with each seminar. If musical events (such as live concerts) are presented before or after a seminar, the SOCIETY's concert rate schedule shall apply.

If video is used in conjunction with mechanical music, the rates outlined above are increased by fifty percent.

STATEMENT OF LICENSEE'S OPERATING POLICY - SEMINARS

Calender Year: ______________________

Number of Seminar Days @ $15: ______________________

$25: ______________________

$35: ______________________

$50: ______________________

Annual License Fee
Based On Above Policy $ ______________________

CERTIFICATE

I hereby certify that the foregoing Statement Of Operating Policy is true and correct as of this ________ day of ________________, 19____.

____
LICENSEE

BY: ______________________

Source: ASCAP, used with permission.

1. The licensing of copyrighted musical compositions for use on commercial records, tapes, computer chips and CDs to be distributed to the public for private use ("mechanical licensing").
2. The licensing of copyrighted musical compositions for use ("synchronization") in audio/visual works including motion pictures, broadcast and cable television programs, CD videos and home videograms.
3. The licensing of copyrighted musical compositions for use in broadcast commercial advertising.
4. The licensing of musical compositions in recordings for other than private use, such as background music, in-flight music, computer chips and syndicated radio services.

The recorded use of music in combination with visual images ("synchronization") does not come within the scope of the compulsory license provisions of the Copyright Act. Such licenses for use of music in films, on TV and in conjunction with home videos must therefore be negotiated on an individual basis between the copyright owner and the prospective user. Agencies like HFA act as an intermediary between the publishers and producers.

After they agree to terms, the agency issues a synchronization license, collects and distributes royalties and periodically monitors users to ensure that the scope of the license issued is not exceeded by unauthorized uses.

The Synchronization Licensing Process

The information required for the issuance of a synchronization license is not complicated. The following will expedite a quotation and licensing:

- The name of the company to whom the license is to be issued
- The title and writers of the composition (and publisher, if known)
- The name of the program, film or video project
- The duration of the use (min./sec.)
- The nature of the use (e.g., background)
- The geographic scope of the use
- The term of license requested (i.e., one year, etc.)

Stock Music

The least complicated source of music for corporate media presentations is stock music.

Some people think that stock music is not copyrighted, but this is incorrect. During the normal course of producing a corporate presentation, the production house that provides the stock music also handles the payment of the license fee to the publisher of the stock music, which, in turn, pays the composer of the music.

Each time a piece of music is used in a corporate media presentation, the producer pays a "drop fee." In other words, the sound production house or stock music house has a set schedule of fees per "needle drop." The producer pays not only a search fee during the course of the production, but also a license fee for each piece of music used in the presentation. The actual amount of the fee depends on the existing arrangement between the stock music house and the stock music publisher. The corporate producer pays a fee based on this preexisting arrangement.

Production houses that specialize in audio production and music libraries can be identified in several trade publications. See Appendix III.

WORKING WITH TALENT

When professional (i.e., union) performers (actors and actresses) are used in corporate presentations, either as on-camera spokespeople, characters in a dramatization or in "live" seminars, sales meetings or other public relations events, these performers are paid, *at minimum*, fees that have been negotiated between the Screen Actors Guild (SAG) and the industrial/educational community.

These minimum fees are set forth in an 87-page contract entitled "Producers-Screen Actors Guild 1987 Codified Industrial and Educational Contract." This contract was modified in 1988 under the title "1988 Non-Broadcast Industrial/Educational Contract/Code, Summary of Changes." This contract runs until April 30, 1990. Essentially, the contract sets minimum fees for performers who appear in nonbroadcast presentations.

The contract covers on-camera performers for basic and point-of-purchase presentations. This aspect sets rates for work by one-day players, stunt day players, three-day players, weekly players, singers and on-camera spokespersons.

The contract also covers such things as off-camera (voice-over) players, extras, pension and health contributions, expanded use beyond original category, supplemental use, non-network television use, foreign television, time of payment, late payment damages, prompting devices, casting and auditions, engagement and cancellation, makeup, hairdress, wardrobe, fitting calls, wardrobe allowance, travel time, flight insurance, transportation, overnight locations expenses, work on a Saturday, Sunday or holiday, overtime, looping, script stunts, upgrade of an extra player, reuse of film or videotape, rest periods, meal periods and night work.

Clearly, a corporate producer (whether in-house or out-of-house) must be cognizant of the responsibilities mandated by the SAG contract from the very moment it becomes evident that a corporate presentation will require a professional performer.

Some companies have become signatories to the SAG contract. Other companies, for various reasons, have chosen to work through a "payroll service" to handle the compensation due actors at the conclusion of a production. Of course, in addition to the fees negotiated, and pension and health contributions, the payroll service charges a fee for handling the paperwork. Figures 11.3 and 11.4 show SAG contract details for corporate media programs.

Corporate producers may choose to work with non-union performers. However, virtually all performers "worth their salt" belong to either SAG or the American Federation of Television & Radio Artists and are covered by union contract. Further, it is not wise to mix union with non-union performers.

Nonprofessional Performers

Often nonprofessionals are asked to appear in a corporate media presentation. These nonprofessionals might include corporate employees, employees of companies associated with the organization or on-the-street citizens.

Strictly speaking, the corporate producer should obtain a release from all the nonprofessionals appearing in the presentation. Technically, the nonprofessional could come back to the producing organization and demand payment for the use "of their likeness" in the presentation. Verbal permission may not be adequate. As with a contract with an outside vendor, getting it in writing precludes a multitude of potential problems. Figures 11.5 through 11.8 are sample photographic releases.

PROTECT YOURSELF, TOO

While this chapter has discussed legal protections with outside resources, it is also important to copyright your own work.

Generally speaking, a copyright notice should be placed on all work created by the corporate media department, whether graphic, filmic or electronic in nature. It must be presumed that in addition to an intended audience for the presentation, other audiences will also gain access to the material. The copyright notice should, for the most part, protect the company's investment in the material.

Figure 11.3: SAG Contract for Corporate Media

The Artist Cannot Waive Any Portion of the Union Contract Without Prior Consent of Screen Actors Guild, Inc.

SCREEN ACTORS GUILD
STANDARD EMPLOYMENT CONTRACT
INDUSTRIAL/EDUCATIONAL
FILM or VIDEOTAPE PROGRAMS

PRODUCER COPY

This Agreement made this ________ day of ________, 19 ____

between ______________________, Producer and ______________________, Performer

1. **SERVICES** — Producer engages Performer and Performer agrees to perform services in a program tentatively entitled ____________ ____________ to portray the role of ____________ to be produced on behalf of ____________ (client).

2. **CATEGORY** — Indicate the initial, primary use of the program.
   - ☐ Category I (Industrial/Educational)
   - ☐ Category II (Point of Purchase)

3. **NUMBER OF CLIENTS** — Indicate number of clients for which program will be used.
   - ☐ Single Client
   - ☐ Multiple Clients

4. **TERM** — Performer's employment shall be for the continuous period commencing ____________, 19____ and continuing until completion of photography and recordation of said role. EXCEPTION - (for Day Performers only) Performer may be dismissed and recalled once without payment for intervening period providing such period exceeds 5 calendar days and Performer is informed of firm recall date at time of engagement. If applicable to this contract, Performer's firm recall date is ____________, 19 ____.

5. **COMPENSATION** — Producer employs Performer as: ___ On-Camera ___ Off-Camera ___ On-Camera Narrator/Spokesperson
   - ☐ Day Performer ☐ Singer, Solo/Duo ☐ General Extra Player
   - ☐ 3-Day Performer ☐ Singer, Group ☐ Special Ability Extra Player
   - ☐ Weekly Performer ☐ Singer, Step Out ☐ Silent Bit Extra Player

   at the salary of { on-camera $ ________ per ☐ DAY ☐ 3-DAYS ☐ WEEK
   off-camera $ ________ for first hour, $ ________ for each additional hour }

6. **OVERTIME** — All overtime rates MUST be computed on Performer's full contractual rate, up to permitted ceilings (NO CREDITING).
   Straight time rate is 1/8th of Day Performer's Rate, 1/24th of 3-Day Performer's Rate, 1/40th of Weekly Performer's Rate.
   Time and one-half rate, payable per hour (1.5 × straight time rate)
   Doubletime rate, payable per hour (2 × straight time rate)
   See section 50 of the contract for details of Weekly and 3-Day Performer for time and one-half and doubletime rates per hour.

7. **WEEKLY CONVERSION RATE** — See section 49 of the contract for details (Day Performers or 3-Day Performers Only).
   The Performer's weekly conversion rate is $ ________ per week.

8. **PAYMENT** — Performer's payment shall be sent to or c/o ______________________.
   Producer must mail payment not later than twelve (12) days (exclusive of Saturdays, Sundays, and holidays) after the day(s) of employment.

9. **ADDITIONAL COMPENSATION FOR SUPPLEMENTAL USE** — Producer may acquire the following supplemental use rights by the payment of the indicated amounts. (Check appropriate items below.) See Section 30 of contract for details of payment.

			Within 90 Days (Total Applic./Actual Salary)	Beyond 90 Days (Total Applic./Actual Salary)
☐	A.	Cable Television		
		Basic Cable (3 years)	15%*	65%*
		Pay Cable (1 year)	20%*	70%*
☐	B.	Non-Network Television, Unlimited Runs	75%	125%
☐	C.	Theatrical Exhibition, Unlimited Runs	100%	150%
☐	D.	Foreign Television - Unlimited Runs outside U.S. & Canada	25%	75%
☐	E.	Non-Network TV, Theatrical & Foreign TV rights as a "Package"	150%	Not Available
☐	F.	Network Television (available **only** by negotiation with and approval of Screen Actors Guild)		
☐	G.	Program for Government Service. Producer may acquire non-network television, theatrical and foreign television rights by payment of 40% of Performer's total applicable salary within 90 days of the completion of principal photography		

* % of total actual salary

10. **WARDROBE** — If PRINCIPAL PERFORMER furnishes own wardrobe, the following fees shall apply for each two-day period or portion thereof:
    Ordinary Wardrobe $ ________ ($12.50 minimum); Evening or Formal Wardrobe $ ________ ($20.00 minimum)
    For **Extra Players** wardrobe fees, please see contract.

11. **SPECIAL PROVISIONS** —

12. **GENERAL** - All terms and conditions of the Producer-Screen Actors Guild Industrial/Educational Contract, as amended, supplemented or re-codified, shall be applicable to such employment.

Producer ______________________ (signature)
Performer ______________________ (signature - (if minor, parent's or guardian's signature))

by ______________________ (Name and Title)
Soc. Sec. ______________________

Address ______________________
Address ______________________

Source: Screen Actors Guild. Used with permission.

Figure 11.4: Summary of Changes in SAG Non-Broadcast Contract

1988 Non-Broadcast, Industrial/Educational Contract/Code

Summary of Changes

1. TERM
The term of the agreement is 19 months: October 1, 1988—April 30, 1990.

2. WAGES
7-1/2% increase on all rates.

Examples

	Category I (Basic)	Category II (Point of Purchase)
DAY PLAYER		
On-Camera	$319.00	$396.00
Narrator/Spokesperson		
First Day	580.00	686.00
Each Additional Day	319.00	396.00
Off-Camera (Voice Over)		
First Hour	$261.00	$290.00
Each Additional 1/2 Hour	76.00	76.00
SINGERS		
On-Camera, per day		
Solo/Duo	319.00	396.00
Group	192.00	237.00
Off-Camera, per hour		
Solo/Duo	171.00	192.00
Group	114.00	128.00

	Category I or II
GENERAL EXTRA	
Per Day	$106.00
AUDIO TAPES (AFTRA Only)	
Principal Performer—First Hour	261.00
Each Additional 1/2 Hour	76.00

3. WARDROBE FEES
Increase the fees in the SAG contract to:

non-evening wear	$15.00
evening wear	25.00

NOTE: The fees in the 1987 AFTRA Code are at this level.

Figure 11.4: Summary of Charges in SAG Non-Broadcast Contract (cont.)

4. **AFTRA HEALTH AND RETIREMENT AND SAG PENSION AND HEALTH**
   Increase contribution rate from 11% to 11-1/2%.

5. **PREFERENCE OF EMPLOYMENT—ORLANDO, FLORIDA AREA**
   Add Preference of Employment zone within a 75 mile radius of Kissimmee, Florida.

6. **EXPANDED USE FOR CATEGORY I PROGRAMS**
   For Category II use of Category I Programs change the rate as follows:
   Within 90 days—50% of applicable salary
   (formerly 100%)
   Beyond 90 days—100% of applicable salary
   (formerly 150%)
   Category I use rights are included in payment for Category II programs.

7. **SUPPLEMENTAL USE**
   Integration and/or Customization—New Provision
   Within 90 days—100% of total applicable salary
   Beyond 90 days—150% of total applicable salary
   Rights for sale and/or rental to industry included in payment of Integration/Customization fee.

   Sale and/or Rental to Industry—New Provision
   Within 90 days—15% of total applicable salary
   Beyond 90 days—25% of total applicable salary

   Supplemental Package Rate
   The rights described above, plus domestic Non-Network Television, Theatrical Exhibition and Foreign Television may be acquired as a package within 90 days of the session by payment of 200% of the total applicable salary. Previously the rate was 150% for domestic Non-Network TV, Theatrical Exhibition and Foreign TV use.

8. **RELEASE TO PAY CABLE, NETWORK TELEVISION AND OVER THE COUNTER SALES**
   In the 1987 agreements, Pay Cable was a Supplemental Use (at 20% of actual salary). In the new agreements only "filler" programs of less than 10 minutes would be covered as a Supplemental release to Pay Cable. All other releases to Pay Cable, and all releases to Network Television and Over the Counter (Cassette) sales would be subject to prior bargaining and agreement with the Union.

9. **POLICY OF NON-DISCRIMINATION AND AFFIRMATIVE ACTION**
   Stronger requirements, bringing this agreement in line with the Commercials Contract, include the following:

Figure 11.4: Summary of Charges in SAG Non-Broadcast Contract (cont.)

- Addition of "Affirmative Action" to the title
- Specific references to "seniors" and "performers with disabilities"
- Affirmative action language "to seek out" has been added, requiring Producers to make good faith efforts to search for performers who are within the protected groups
- Producer shall not use any information contained on the INS Form I-9 to discriminate against any performer on the basis of sex, race, age, or national origin. Such information shall be maintained in confidence

10. MINORS
All calls for fittings, tests and interviews must occur after school hours and be completed before 7 P.M.

11. TRAVEL TIME
Elimination of 30 minutes free travel time, to and from the set, on overnight locations.

12. HOLIDAYS
Add Martin Luther King, Jr.'s birthday.

13. WORK IN SMOKE
Notification of work in smoke required prior to booking.

Source: Screen Actors Guild. Used with permission.

Figure 11.5: Sample Photography Release

Property Release

For good and valuable consideration herein acknowledged as received, the undersigned being the legal owner of, or having the right to permit the taking and use of the photographs of certain property designated as ____________________
does grant to ____________________,
his agents or assigns, the full rights to use such photographs and copyright same, in advertising, trade or for any purpose.

I also consent to the use of any printed matter in conjunction therewith.

I hereby waive any right that I may have to inspect or approve the finished product or products or the advertising copy or printed matter that may be used in connection therewith or the use to which it may be applied.

I hereby release, discharge and agree to save harmless, ____________________,
his legal representatives or assigns, and all persons acting under his permission or authority or those for whom he is acting, from any liability by virtue of any blurring, distortion, alteration, optical illusion, or use in composite form, whether intentional or otherwise, that may occur or be produced in the taking of said picture or in any subsequent processing thereof, as well as any publication thereof even though it may subject me to ridicule, scandal, reproach, scorn and indignity.

I hereby warrant that I am of full age and have every right to contract in my own name in the above regard. I state further that I have read the above authorization, release and agreement, prior to its execution, and that I am fully familiar with the contents thereof.

Dated: ____________________ ____________________

____________________ ____________________

Witness

____

(Address)

Note: This release can be used for film, video and other media.
Source: American Society of Magazine Photographers. Used with permission.

Figure 11.6: Sample Model Release for Adult

Adult Release

In consideration of my engagement as a model, and for other good and valuable consideration herein acknowledged as received, upon the terms hereinafter stated, I hereby grant ______________________________, his legal representatives and assigns, those for whom ______________________________ is acting, and those acting with his authority and permission, the absolute right and permission to copyright and use, re-use and publish, and republish photographic portraits or pictures of me or in which I may be included, in whole or in part, or composite or distorted in character or form, without restriction as to changes or alterations, from time to time, in conjunction with my own or a fictitious name, or reproductions thereof in color or otherwise made through any media at his studios or elsewhere for art, advertising, trade, or any other purpose whatsoever.

I also consent to the use of any printed matter in conjunction therewith.

I hereby waive any right that I may have to inspect or approve the finished product or products or the advertising copy or printed matter that may be used in connection therewith or the use to which it may be applied.

I hereby release, discharge and agree to save harmless ______________________________, his legal representatives or assigns, and all persons acting under his permission or authority or those for whom he is acting, from any liability by virtue of any blurring, distortion, alteration, optical illusion, or use in composite form, whether intentional or otherwise, that may occur or be produced in the taking of said picture or in any subsequent processing thereof, as well as any publication thereof even though it may subject me to ridicule, scandal, reproach, scorn and indignity.

I hereby warrant that I am of full age and have every right to contract in my own name in the above regard. I state further that I have read the above authorization, release and agreement, prior to its execution, and that I am fully familiar with the contents thereof.

Dated: ______________________________

____

____

____
(Address)

____
(Witness)

Source: American Society of Magazine Photographers. Used with permission.

Figure 11.7: Simplified Model Release

Simplified Adult Release

Dated ______________________________

For valuable consideration received, I hereby give __________________________ the absolute and irrevocable right and permission, with respect to the photographs that he has taken of me or in which I may be included with others:

(a) To copyright the same in his own name or any other name that he may choose.

(b) To use, re-use, publish and re-publish the same in whole or in part, individually or in conjunction with other photographs, in any medium and for any purpose whatsoever, including (but not by way of limitation) illustration, promotion and advertising and trade, and

(c) To use my name in connection therewith if he so chooses.

I hereby release and discharge ________________________ from any and all claims and demands arising out of or in connection with the use of the photographs, including any and all claims for libel.

This authorization and release shall also enure to the benefit of the legal representatives, licensees and assigns of ______________________________as well as, the person(s) for whom he took the photographs.

I am over the age of twenty-one. I have read the foregoing and fully understand the contents thereof.

____

____
Witnessed by:

Source: American Society of Magazine Photographers. Used with permission.

Figure 11.8: Sample Model Release for a Minor

Minor Release

In consideration of the engagement as a model of the minor named below, and for other good and valuable consideration herein acknowledged as received, upon the terms hereinafter stated, I hereby grant ______________________________, his legal representatives and assigns, those for whom ______________________________ is acting, and those acting with his authority and permission, the absolute right and permission to copyright and use, re-use and publish, and republish photographic portraits or pictures of the minor or in which the minor may be included, in whole or in part, or composite or distorted in character or form, without restriction as to changes or alterations from time to time, in conjunction with the minor's own or a fictitious name, or reproductions thereof in color or otherwise made through any media at his studios or elsewhere for art, advertising, trade or any other purpose whatsoever.

I also consent to the use of any printed matter in conjunction therewith.

I hereby waive any right that I or the minor may have to inspect or approve the finished product or products or the advertising copy or printed matter that may be used in connection therewith or the use to which it may be applied.

I hereby release, discharge and agree to save harmless ______________________________, his legal representatives or assigns, and all persons acting under his permission or authority or those for whom he is acting, from any liability by virtue of any blurring, distortion, alteration, optical illusion, or use in composite form, whether intentional or otherwise, that may occur or be produced in the taking of said picture or in any subsequent processing thereof, as well as any publication thereof even though it may subject the minor to ridicule, scandal, reproach, scorn and indignity.

I hereby warrant that I am of full age and have every right to contract for the minor in the above regard. I state further that I have read the above authorization, release and agreement, prior to its execution, and that I am fully familiar with the contents thereof.

Dated: ______________________________

______________________________	______________________________
(Minor's Name)	(Father) (Mother) (Guardian)
______________________________	______________________________
______________________________	______________________________
(Minor's Address)	(Address)

____

(Witness)

Source: American Society of Magazine Photographers. Used with permission.

One of the most important factors is to make sure a copyright legend is included on all work.

The ideal copyright legend is as follows:

"Copyright © by XYZ Corporation 198—
All Rights Reserved"

(The use of the words, "All Rights Reserved" places the work within the copyright protection of the Buenos Aires Convention, one of the international copyright treaties.)

It is a criminal offense to infringe a copyright willfully and for purposes of commercial advantage or financial gain. It is also a criminal offense to affix a fraudulent copyright notice, to remove a copyright notice or to make a false statement in a copyright registration application.

The Copyright Office in Washington, DC is primarily a place of record where claimants may register their copyright materials. (It does not furnish legal advice except to explain the copyright law itself as well as to report on various facts which it has on record.) In order to sue someone for copyright infringement, the corporation must first register work in the Copyright Office by completing an application and sending two copies of the work together with a filing fee.

SUMMARY

The corporate media manager and/or producer should have a sign behind his or her desk that says "Take Nothing For Granted."

This is especially true with respect to contracts, copyrights and permissions. If there is any doubt as to the contract between the corporation and an external resource, the copyright nature of a graphic, filmic or electronic work, or whether permission needs to be garnered from a professional or nonprofessional performer, the manager and/or producer should leap to action.

The need to be cognizant of the legal aspects of corporate media work is especially important today and will become even more of an issue for the future. The reason is that new distribution channels for corporate media presentations have evolved. Each year, more households have cable television and VCRs. Satellite technology has become commonplace.

As a result, corporations are increasingly using these new modes of distribution for both internal and external corporate messages, particularly marketing and public relations. This impacts directly on the potential problems a company could face if legal consideration such as contracts, copyrights and permissions are ignored. There is a difference, for example, between a fee paid to a performer for an on-camera appearance in a corporate marketing piece shown at a trade show and the same piece aired on a cable television channel.

Technology is changing the distribution landscape. The legal aspects of corporate media production work have, therefore, become more important to the corporate media producer.

12

The Evolving Corporate Media Production Function

What we define as the role and scope of the corporate media production function of today may well change in the next decade. In an era of social and technological change, media managers must be aware of both emerging communications technologies and social trends, and how they will change the nature of the corporate media operation. This chapter gives an overview of some of the new technologies and trends, and tries to judge just how the corporate media department will adjust to these changes.

TRENDS IN COMMUNICATIONS TECHNOLOGY

The Information Society

We are experiencing a change in the way we "transport" information. In a broad sense, we could say that the car, the train and the plane transport information physically. However, information can also be transported "electronically." What is emerging is a shift from an era of physical transportation of information to an era of electronic transportation. James Martin described this in *The Wired Society*, where multinational corporations are:

> ...Laced together with worldwide networks for telephones, instant mail and links between computers. Video conference rooms and computerized information systems increase the degree to which head-office executives guide corporate operations in other countries. Computers schedule fleets and optimize the use of resources on a worldwide basis. Money can be moved electronically from one country to another and switched to different currencies. There is worldwide management of capital, inventory control, product design, bulk purchasing, computer software, and so on. Local problem situations can trigger the instant attention of head-office staff.[1]

Computers and Integrated Circuits

The scenario painted by Martin is in many ways a present reality. Nearly everyone today has a pocket calculator, and sales of home computers are increasing. The growth in the number and kind of electronic devices is due largely to the development of two interactive devices: the computer and integrated circuits. The first electronic computer, ENIAC (for Electronic Numerical Integrator and Calculator), was built in 1946. It consumed 140,000 watts of electricity and contained 18,000 vacuum tubes. In 1947 Bell Laboratories introduced the transistor, a tiny piece of semiconducting material, such as silicon or germanium. The transistor was the perfect mate for digital computers using a binary code. In 1959 Texas Instruments and Fairchild Semiconductor simultaneously announced the production of integrated circuits: single semiconductor chips containing several complete electronic circuits. By 1970 laboratories were producing chips with large scale integration of circuitry (LSI); thousands of integrated circuits crammed onto a single quarter inch of silicon.

Today more computer power has been put into less and less space. There are now electronic devices that create graphics. General Electric developed electronic graphic devices for simulation purposes for the National Aeronautics and Space Administration (NASA) space program. Such devices (called genigraphics) are now used for the generation of graphics and the ultimate transfer of these electronic images to 35mm slide format.

Computer devices and integrated circuitry have invaded other aspects of the communications world, for example, with devices that program and run multi-image presentations. Computer-like devices also perform functions for videotape editing, color correction of film-to-tape transfers, and "timing" of film prints. All of these devices have computer-like mechanisms and integrated circuitry, as well as a keyboard terminal which allows the operator to manipulate the "information" in the device.

Cable Television

Cable television was born out of a need to bring television signals to communities that could not receive good broadcast signals. Cable television distributes radio-frequency television signals to subscribers' TV sets via a coaxial cable and/or optical fiber, rather than by broadcasting the signal over the airways. The cable, therefore, provides interference-free signals; moreover, whereas telephone lines are not presently able to carry the kind of video signals that coaxial or fiber optic cables can carry, the reverse is true.

Cable television's impact on the broadcasting industry seems to have reached a critical mass in the early 1980s. According to the National Cable Television Association, cable penetration of television households in the United States (with respect to basic cable) reached 25.3% by February 1981. By November 1986 penetration was 48.1%. As of this writing penetration is more than 50%.

Satellites

Satellite technology, together with the growth of cable systems, allows a programmer to direct electronic content at specific regions on demand. It creates the possibility of a geometric increase in the number of programming possibilities.

Satellite technology is creating the potential for the development of "electronic highways" heretofore not existent. New companies have been created for the sole purpose of providing business communications, including voice communications, data processing, facsimile transmissions and teleconferencing.

Teleconferencing and business television networks are becoming more commonplace, as the cost of systems decreases and the quality rises. Further:

> Today, satellites transfer most long-distance telephone calls, beam prime time network TV programs to affiliates nationwide, link regional offices of multi-national businesses with corporate headquarters, and provide access to more than 100 channels of TV programming to thousands of cable TV affiliates and millions of home satellite TV owners.[2]

Home Video

The home video market also reached critical mass in the early 1980s.

By the end of 1981, videocassette recorders (VCRs) were in 26% of all television households in the U.S. In 1988, VCRs had achieved 60% penetration.

Videodisc

Another electronic technology that is becoming a reality is the videodisc. Videodisc technology represents "the merging of laser technology, computer technology and information storage technology."[3] We are not merely referring to the simple playback (or capacitance) videodisc system, but to the interactive (laser optical) videodisc that incorporates a microprocessor accessed by a hand-held device. This interactive capability allows the user to locate any frame on the videodisc (or any part of the program) without time-consuming searching. Some of the most significant aspects and major advantages of optical videodisc technology are:

- The videodisc resembles the long-playing record, except that it carries both images and sound. The videodisc player is connected to a standard television set on which the image is viewed.
- The optical videodisc has freeze-frame, slow motion and reverse capability, which makes it well-suited for educational and industrial applications.
- The videodisc does not have to be duplicated in real time—a disc with two hours of programming can be duplicated in less than one minute.
- The optical videodisc utilizes light from a laser for playback. Since there is no physical contact with the surface of the disc, it can be used without wear. Furthermore, it has a protective coating so that dust, fingerprints or scratches have virtually no effect on picture quality.
- Each side of the disc can carry 54,000 frames or pictures, each of which can be randomly accessed by a hand-held, programmable remote control unit. Thus an analog videodisc containing the equivalent of 60 billion bits of data per disc side can be scanned in seconds.
- The optical videodisc has two sound tracks which allow for stereophonic sound, or tracks in two different languages, or levels of instruction.[4]

The interactive videodisc could have a major impact on the way information is learned, since it combines several elements of instruction: printed text, audiovisual presentation, color and individualized instruction.

CD-ROM, Desktop Video and HDTV

Other electronic technologies are making their way into the organizational media marketplace. Several companies have begun to market computer CDs, known as CD-ROM (for "read only memory"). They are only 4.72 inches in diameter but can store as much information as a stack of typewritten pages nine stories high (*Time*, April 11, 1988, p. 52). The newest discs take advantage of the medium's vast capacity for storing pictures and sounds as well as words. A CD device can come with a built-in computer and can be hooked up to a television set. With a hand-held controller the user could interact with the televised information.

Several companies are now marketing desktop video devices that are to video animation what desktop publishing is to printing. For less than $20,000, the price of the software and hardware, companies can input all manner of graphic materials, create animation and present the finished "video" to employees or potential customers within a matter of hours (*Business Week*, June 1, 1987, p. 85).

High Definition Television is making its way into the marketplace. Several European companies have already developed the technology to adapt current American standards to a high-definition look. Some feature films and commercials are being shot in HDTV and "down converted" to meet the lower quality American television standards. It is only a matter of time before the United States ultimately converts to HDTV, perhaps in concert with a worldwide conversion to a one-standard television system.

THE EVOLVING SOCIETY

Demassification of Society

John Naisbitt, former Senior Vice President of Yankelovich, Skelly and White, has noted certain trends which he feels point to a "demassification" of society—trends that will have an impact on the organizational environment in which the media department operates, notably:

- The United States is rapidly shifting from a mass industrial society to an information society.
- There is more decentralization than centralization taking place in America.
- There are the beginnings of a job revolution in America, a basic restructuring of the work environment from top-down to bottom-up.[5]

Essentially, these trends are indicative of changes in the pattern and flow of communication.

Employee Motivation

David Rockefeller, when Chairman of the Chase Manhattan Bank, observed that the increasing numbers of college and business school graduates will soon swell the ranks of middle and upper management, leaving many individuals unable to find positions commensurate with their training—a situation already in evidence today. Motivating these employees, he feels, will pose a "supreme test" for the manager of the future.[6]

In addition, Alvin Toffler describes the workers of the Third Wave, the "new workers:"

> Workers are forced to cope with more frequent changes in their tasks and with a blinding succession of personnel transfers, product changes and reorganizations.
>
> What Third Wave employers increasingly need, therefore, are men and women who accept responsibility, who understand how their work dovetails with others', who can handle ever larger tasks, who adapt swiftly to changed circumstances and are sensitively tuned in to the people around them.[7]

Corporate media production employees and managers, who are so much a part of communications advances, will be in the vanguard of the Third Wave.

Globalization

An even more important societal factor is the continued trend towards globalization, particularly economic globalization.

Ever since the October 1987 stock market "crash" there can be no doubt that national economies are increasingly connected in a global economy. In this writer's opinion, the evolving world economy is a direct result of the increasing use of electronic devices. What we are observing, I believe, is the further realization of what Marshall McLuhan called the "global village" in his 1964 work *Understanding Media: The Extensions of Man.*

In coining this phrase, McLuhan was referring to the globalization of communications (through such electronic technologies as satellites, radio and television) to the point where societies could communicate with such speed (as in the speed of light) that the effect would place all of us in the context of a village, in this case a "global village."

Recent trends in marketing tend to confirm McLuhan's observation. In 1992 the European Common Market intends to become, in effect, one market. Increasingly, advertising agencies are becoming global in organization (with the British in the lead). American companies are merging, particularly consumer products companies, to take advantage of global marketing opportunities. Broadcast and cable television programmers are increasingly extending their global distribution. Banks and other financial service companies are looking at worldwide markets.

All of these trends are based on the exponential opportunities created by electronic technologies. This generation may well be witnessing the same kind of so-called "revolution" created when the Greeks developed their alphabet circa 700 B.C., Gutenberg developed his printing press in 1455, and the first successful electric telegraph line was introduced in America in 1844.

What we may be witnessing is the total restructuring of all the economies on this earth, a massive shift in what countries create and move information, what countries create food, what countries manufacture products. This worldwide "happening" is a direct result of the evolution of electronic technologies. And it is a phenomenon that is as irrepressible as a glacier in an Ice Age. It is inexorable.

IMPLICATIONS FOR THE FUTURE

The developments reviewed in the preceding pages lead to certain general conclusions, and possible implications, for the future.

(1) There is a trend away from homogeneity and toward heterogeneity. Programming for, and marketing to, mass audiences may give way to regional and even local programming; "narrowcasting" is a concept already much in evidence: home video, cable systems and satellites will help foster this trend. This may result in regional or local marketing or public relations programming being produced by some media production department directly. In addition, the impact of burgeoning cable systems, together with satellites, augurs a change in the function of existing broadcasting systems as we know them. The new electronic highways will enable programmers to bypass the usual distribution systems.

(2) External communications activities will change. The focus on local or regional politics will cause organizations to modify their government relations activities.

(3) Internal communications activities will also change. More and more workers will be employed in information processing activities, as opposed to "manufacturing" activities. There will be better educated employees, including more women and minorities, striving for higher

levels of achievement. This will cause stresses in many organizations which will, in turn, change the configuration of internal organizational communications systems: they may have to become more personal and frequent because employees will require more contact with higher management, as well as more independent decision-making opportunities.

(4) Organizations will tend to be "flatter." Layers of middle management will be eliminated. Management span of control will be extended. Federal Express will become an organizational model to be emulated.

(5) Electronic technology will continue to march forward, with more and more electronics merging into more and more devices. Increasing numbers of electronic devices are interactive, and this trend will continue. Furthermore, the information storage capacity of such devices is on the rise, particularly with respect to the computer and the videodisc.

(6) Electronic devices will become less expensive for the recording, manipulation, editing and retrieval of information, while the more traditional information-carrying devices, such as paper and film, will be used less as a news or even as an entertainment medium. The decrease in the use of film for news programming and the increase in the use of ENG-type video cameras and recording equipment is one example of the probable demise of film as a corporate communications technology.

(7) As the cost for disseminating information electronically decreases and its efficiency as a communications medium increases, there will be a decrease in the use of paper as a "mass" communications device. Increased costs of mailing and of producing paper, and the unwieldiness of information in print form as compared to its electronic equivalent, will contribute to this decline. However, this does not mean that paper will be abandoned as an information-bearing technology. On the contrary, paper will have other functions; for example, as the hardcopy form of information retrieved electronically.

(8) The growth of electronic devices in the home, in particular home videocassette recorders, will usher in an era in which organizations produce programming that will be viewed by employees and consumers in the home.

(9) Organizational use of electronic devices will increase as the need for higher employee productivity increases. The increased use of elec-

tronic devices for word processing and data processing is but one example.

(10) More manufacturing operations will be handled by machines run by computers. As a result, a lower percentage of workers will be engaged in manufacturing work as machines do more manual labor.

Implications for the Media Operation

These fundamental changes in the larger societal and technological landscape will have a direct impact on various aspects of the corporate media production operation.

Media Product

The output of organizational media production departments will probably increase as the pace of change and the pressures of internal and external communications activities increase.

Technological changes, especially in administrative, manufacturing and communications systems, will create the need for informational and training programs. For example, as more and more organizations switch over to electronic word processing, training programs will have to be created to train both users and administrators. As increasing numbers of organizations use machines to direct manufacturing operations, the need for programming to inform and train personnel on the use, operation and maintenance of these machines will become greater. As organizations use more sophisticated electronic systems for communications, the need for information and training programs on these systems will increase. Furthermore, the process will be self-generating as newer and more capable electronic systems are brought into use.

The advent of electronic devices will intensify the need for increased communications activities among various components of the organization. This will be especially true of the larger, far-flung multinational organizations. While this does not mean a tendency towards centralization from a communications or control point of view, it does mean that in a changing external environment, organizations will require increased communications to maintain equilibrium. The need to create understanding among the various organizational components will become imperative to organizational survival. This will again translate into an increased need for communications product.

The need for more communications activity together with the rising cost of travel will cause greater organizational use of teleconferencing systems. Together with the increased use of satellites for such purposes, teleconferencing may reduce business travel to a noticeable degree.

The development of cable systems in all their various configurations, fostered by the increased use of satellites, could have an impact on the marketing and advertising of organizational products. This in turn will influence the nature, form and origination of marketing and advertising programming. It is conceivable that via a combination of cable systems, organizations will develop regional and even local marketing and advertising programming. This type of audience segmentation may result in more professional in-house media operations becoming involved in the development, production and distribution of such programming.

Marketing information directed at organizational sales forces should also increase as competition, both domestic and foreign, intensifies. Management communications programming will also intensify as central management must communicate more frequently with field managers.

This general increase in the use of corporate media capabilities, as a result of the increased need for communications, will also increase the use of external resources.

Hardware Systems

Organizational media departments will experience inevitable changes in hardware systems. As electronics continues to pervade the culture, we should see, for example, the further development and use of computer-like devices for the creation of graphics. Similar devices will be used for the creation and manufacture of 35mm slides. Moreover, it is possible that the need for hardcopy graphics in the form of 35mm slides will decrease as other kinds of computer-like devices become accessible to managers, so that data, statistics and trends can be created electronically, even up to the last minute before a presentation.

Video cameras will contain more electronic control devices; even tubes may disappear. Cameras will become smaller, more sophisticated, capable and cheaper. Devices for electronic editing will become more flexible, sophisticated and less expensive. Videodisc technology may move toward the development of devices that both play back and record video information. If this happens on a level that can be afforded by most, it is conceivable that videotape may drop out as a communications technology, just as it appears that film will. Moreover, the further development of large screen video projection may answer the need for large audience viewing of programs originated and produced on videotape or videodisc.

Thus, by the 1990s, organizations may no longer be weighing the merits of filmstrips versus film or videotape, but rather weighing the merits of videodisc versus videotape—filmstrips, slide/tape, Super 8mm, and 16mm film distribution having vanished almost altogether.

Organizations will also face the possibility of installing teleconferencing facilities and satellite up/down links as communications via these technologies increase in frequency. By the 1990s, teleconferencing may be as casual a communications activity among organizational management as picking up the telephone to make a transcontinental phone call is today.

Organizational Changes

There will also be a converging of technologies from other parts of the organization. The first and most obvious convergence should be with regard to teleconferencing. Organizational media operations will "bump" into hardware systems operated by the telecommunications department, which, in turn, will bump into technologies operated by the electronic data processing department. Administrative operations, such as word processing and electronic mail, may also converge. The reason for this increased interaction among seemingly disparate organizational departments is that teleconferencing technologies, particularly via satellite communications, can handle all the various communications activities: voice communications, data processing, facsimile transmission and two-way, wide-band video communications.

In Chapter 3, an "ideal" model for the organization of the media operation was presented. While some organizations more or less reflect this kind of model, as of this writing the model does not represent the vast majority. The tendency toward centralization of media activities will come as a result of two trends: an organizational focus on increased productivity, and the increased use of electronic technologies.

As organizations move toward maximizing resources in a world of limited resources, they will look for ways to make the best use of existing resources. This may mean bringing seemingly disparate activities together and managing these resources under one organizational umbrella. Furthermore, as more electronic devices are used in media activities, we should see a continued shift away from the manual creation of communications products, such as graphic arts and photography, to the electronic creation of such products. Thus, by the 1990s, media production activities, whether graphic, photographic or electronic, will have electronic/computer technologies in common. In turn, it should become obvious these production activities belong together.

As media activities become more electronic and computerized, it is conceivable these activities will have more in common with telecommunications, data processing and word processing as opposed to marketing, training or corporate communications. It is possible, therefore, that the media department will move away, organizationally, from its traditional home, and toward a new area with which the media department will have more in common—technology.

On the broader scale, the organizational need to better use communications systems and better control overall organizational communications activities may result in the centralizing of other communications functions. It is possible that by the next decade, more organizations will have merged their training, employee communications, advertising and marketing, press relations, government relations, community relations, media production department and other "communications systems," such as telecommunications, into one "macro" communications activity. Such an activity might be divided into three major areas:

- Internal communications: training, employee communications.
- External communications: advertising/marketing, press relations, government relations, community relations, investor relations.
- Media operations: graphic arts, film, videotape, videodisc, teleconferencing, telecommunications, word processing, data processing.

In this context, both internal and external communications activities are "audience" oriented, whereas the media operation is software and hardware oriented. Media staffs would have the overall function of translating communications content into a media form (software) and using various technologies (hardware) to deliver that content to the appropriate audience.

Implications for the Media Production Staff

The type of expertise found among the media staff will change. The further development of videotape technologies (cameras, recorders and editing equipment) will necessitate a great deal of upgrading of the skills of technicians and producer/directors. While manufacturer advertisements give the impression that the newer videotape technologies make production and post-production a lot more accessible, it must be remembered that maintenance becomes more complex. Thus, organizations can expect an increase in the number of personnel involved in media activities, especially in the technical area. Moreover, existing and future personnel will be expected to have a higher degree of understanding of the communications problem-solving process, existing and developing technologies, and organizational practices.

The videodisc, especially, will change the nature of media staff expertise. Whereas programming for slide, film and videotape is linear in presentation, programming for the videodisc is not. Moreover, videodisc programming will present opportunities for the use of still graphics, as well as motion graphics (such as film and videotape). When to use what medium for informational or instructional purposes will become of paramount importance in the effectiveness of videodisc programming. As the use of videodisc increases, media production employees (especially producer/directors) should find themselves in much greater contact with training personnel who, presumably, are the experts in instructional design. The videodisc, because of its high interactive capability, will do much to bring together media and training department staffs. This should foster even greater cooperation among the two professional groups, which will have a net positive effect on the organization.

All personnel in the media production department will be affected by the electronic technologies. Graphic artists will find themselves designing and executing graphics at an electronic keyboard rather than at a drawing table, using a television-like screen to make changes in design, colors and hue, both for pictorial representations and word copy. Computer-graphics devices may have a substantial impact on the sale of drawing implements.

Photographers, too, will be affected by the advent of electronic technology. The creation of 35mm slides has already been affected by the development of devices that create graphics electronically and then convert them to either 35mm slide or paper hardcopy. Motion film (Super 8mm and 16mm) may very well see its demise in the 1990s, as videotape and videodisc recording devices are used more. Writers, too, will find themselves working at word processing terminals rather than typewriters.

A FINAL WORD

While it appears that many corporate media operations will experience an evolutionary transformation during the 1990s, not all will. Moreover, it is not necessarily true that with the advent of such technologies as computer graphics, teleconferencing and videodisc, suddenly all organizational media departments will become gorged with sophisticated electronic gadgetry and find themselves staffed with electronic media production superstars.

On the contrary, the above scenario may be limited to those media operations with sufficient need and means for such hardware, software and personnel. Yet, many organizations—perhaps all—will be touched by the evolution of electronic technologies. Moreover, new technologies will not completely obsolesce existing technologies. The history of technology teaches us that newer technologies have the effect of changing the function of existing technologies, rather than making them obsolete altogether. While speech was once the primary form of communications, writing and print have not, obviously, done away with it. Photographic and electronic media have not done away with writing and print.

Looking to the future, highly interactive communications media, such as teleconferencing and the videodisc, should not do away with photography and videotape technologies altogether, but will probably change their function.

For example, videotape could change the long-term function of print as an organizational mass communications tool. For example, one impact of videotape on print has been the development of employee news show on video, which transformed many print house organs from a news medium to a feature medium.

Whereas film was once relied upon heavily as a news medium by the broadcast networks and local stations, it has been largely replaced by videotape and electronic news-gathering hardware. However, film continues to be used as an entertainment medium; e.g., for the movies shown in theaters, and on home television systems (either broadcast, pay cable, videocassette or videodisc). However, we should note that more and more movies are being shot on videotape and transferred to the film medium for distribution and presentation, so that film's survival as a production medium may be in question.

The function of videotape may also change in the next decade, especially if videodiscs are developed (at a cheap enough price) that play back *and* record video information. If this occurs—and the technologies are in the process of development now—it could very well change the function of videotape as a communications medium both in consumer markets and organizational life.

Further, while new technologies will continue to appear, this does not mean to say people will readily accept them. In the organizational context, at least, employees tend to resist technological change. It is entirely possible, for example, that middle managers will resist the use of video teleconferencing because it will mean "not getting away from the office." If, indeed, teleconferencing cuts down on business travel, this could be perceived as a loss of status and prestige. Another example is the resistance of managers to the use of keyboard terminals to access information banks, which may be perceived as a technician's function.

Be that as it may, new technologies, especially if they are more effective and efficient than existing technologies, have an historical habit of taking hold anyway. It is not really a matter of "if" new technologies will have a meaningful impact, but "when."

The next five to 10 years will witness change, both socially and technologically. Perhaps the most significant change will be the increasing economic interdependence among nations. Organizational and media management must be prepared for it. Presumably, change is the primary reason for the media department: change creates the need to communicate. Media staffs prepared both technically and attitudinally for change will benefit the most.

The future will present many problems for organizations. This will also serve as an opportunity for media production departments to help organizations through an era that will witness the transformation of our civilization from primarily a "resource processing" (manufacturing) to an "information processing" society on a global scale.

FOOTNOTES

1. Martin, James, *The Wired Society* (Englewood Cliffs, NJ: Prentice-Hall, 1978), p. 13.

2. Long, Mark, *1985 World Satellite Almanac: The Complete Guide to Satellite Transmission and Technology* (Boise, ID: Commtek Publishing Company, 1985).

3. Speech by John J. Reilly, President of DiscoVision, 1980.

4. Roberts, Martin, "Video Disc Capabilities and Effects." Remarks given at the Hearing of the House of Representatives, Government Activities and Transportation Subcommittee of the Committee on Government Operations, Federal Building, Los Angeles, CA, 1979, and Sigel, Efrem, et al., *Video Discs: The Technology, the Applications and the Future* (White Plains, NY: Knowledge Industry Publications, Inc., 1980).

5. Naisbitt, John, "What's Really Happening in the U.S." Speech given before the Foresight Group, Stockholm, Sweden, September 20, 1979.

6. Speech by David Rockefeller. Given before the Commonwealth Club of San Francisco, November 2, 1979.

7. Toffler, Alvin, *The Third Wave* (New York: William Morrow & Co., 1980).

Appendix I: Recommended Books

Abelow, Daniel and Hilpert, Edwin J. *Communications in the Modern Corporate Environment*. Englewood Cliffs, NJ: Prentice-Hall, Inc., 1986.

Barnouw, Eric. *Mass Communications*. New York, NY: Holt, Rinehart and Winston, 1958.

Bass, Bernard M. *Leadership and Performance Beyond Expectations*. New York, NY: The Free Press, 1985.

Bedeian, Arthur G. *Management*. Chicago: The Dryden Press, 1986.

Bellman, Geoffrey M. *The Quest for Staff Leadership*. Glenview, IL: Scott, Foresman and Co., 1986.

Bennis, Warren and Nanus, Burt. *Leaders: The Strategies for Taking Charge*. New York: Harper & Row, 1985.

Bittel, Lester R. *Leadership: The Key to Management Success*. New York: Franklin Watts, 1984.

Blanchard, Kenneth, et al. *Leadership and the One Minute Manager*. New York, NY: William Morrow and Co., 1985.

Bradford, David L. and Cohen, Allan R. *Managing for Excellence*. New York, NY: John Wiley & Sons, 1984.

Brooks, Julie K. and Stevens, Barry A. *How to Write a Successful Business Plan*. New York, NY: Amacom, 1987.

Brown, James W., et al. *AV Instruction: Technology, Media, and Methods*. New York: McGraw-Hill, 1973.

Brown, L. and Walker, S. W. (Eds). *Fast Forward: The New Television and American Society*. Kansas City: Andrews & McMeel, 1983.

Brownstone, David M. and Franck, Irene M. with R. Guiley. *The Manager's Advisor, ED RV*. New York: Amacom, 1987.

Brush, Judith M. and Brush, Douglas P. *Private Television Communications*. Cold Spring, NY: D/J Brush Associates, 1987.

Budd, John F., Jr. *Corporate Video in Focus: A Management Guide to Private TV*. Englewood Cliffs, NJ: Prentice-Hall, Inc., 1983.

Bunyan, John A. *Why Video Works: New Applications for Management*. White Plains, NY: Knowledge Industry Publications, Inc., 1987.

Buzzell, Robert D. and Gale, Bradley T. *The PIMS Principles: Linking Strategy to Performance*. New York: The Free Press, 1987.

Choate, Pat and Linger, J. K. *The High-Flex Society*. New York: Alfred A. Knopf, 1986.

Conover, Theodore E. *Graphic Communications Today*. West Publishing, 1985.

Corporate Culture—Diagnosis and Change. New York: St. Martin's Press, 1986.

Corrado, Frank M. *Media for Managers*. Englewood Cliffs, NJ: Prentice-Hall, Inc., 1984.

Crosby, Philip B. *Running Things: The Art of Making Things Happen*. New York: McGraw-Hill Book Co., 1986.

Crow, Wendell C. *Communication Graphics*. Englewood Cliffs, NJ: Prentice-Hall, Inc., 1986.

Czitrom, Daniel J. *Media and the American Mind from Morse to McLuhan*. The University of North Carolina Press, 1982.

Degen, Clara, ed. *Understanding and Using Video: A Guide for the Organizational Communicator*. New York, NY: Longman, 1985.

Delaney, Robert V., Jr. and Howell, Robert A. *How To Prepare An Effective Business Plan: A Step-By-Step Guide*. New York: Amacom, 1986.

Delaney, William A. *Tricks of the Manager's Trade: How to Solve 30 Common Management Problems*. New York: Amacom, 1982.

Dessler, Gary. *Management Fundamentals: Modern Principles and Practices, Ed. 4*. Reston, VA: Reston Publishing Co., 1985.

Diebold, John. *Making The Future Work*. New York: Simon and Schuster, 1984.

Drucker, Peter F. *The Changing World of the Executive*. New York: Times Books, 1982.

Frost, Peter J., et al. *Organizational Culture*. Beverly Hills, CA: Sage Publications, 1985.

Garland, Ken. *Graphics Handbook*. New York, NY: Van Nostrand Reinhold, 1966.

Gayeski, Diane and Williams, David. *Interactive Media*. Englewood Cliffs, NJ: Prentice-Hall, Inc., 1985.

Gitman, Lawrence J., Joehnk, Michael D., and Pinches, George E. *Managerial Finance*. New York: Harper & Row, 1985.

Goodchild, Jon and Henkin, Bill. *By Design: A Graphics Sourcebook of Materials, Equipment and Services*. New York: Quick Fox, 1980.

Gross, Lynne Schafer. *Telecommunications: An Introduction to Radio, Television, and the Developing Media*. Dubuque, IA: Wm. C. Brown, 1983.

Hatakeyama, Yoshio. *Manager Revolution*. Stamford, CT: Productivity Press, 1985.

Heirs, Ben with Peter Farrell. *The Professional Decision-Thinker*. New York: Dodd, Mead & Company, 1987.

Hickman, Craig R. and Silva, Michael A. *Creating Excellence: Managing Corporate Culture, Strategy and Change*. New York: New American Library, 1984.

Hickman, Craig R. and Silva, Michael A. *The Workbook for Creating Excellence*. New York: New American Library, 1986.

Karen, Ruth, Ed. *Toward the Year 2000*. New York: William Morrow and Co., Inc., 1985.

Kellerman, Barbara. *Leadership: Multidisciplinary Perspectives*. Englewood Cliffs, NJ: Prentice-Hall, Inc., 1984.

Kerr, Clark. *The Future of Industrial Societies: Convergence or Continuing Diversity*. Cambridge, MA: Harvard University Press, 1983.

Lardner, J. *Fast Forward: Hollywood, the Japanese and the VCR Wars*. New York: W. W. Norton & Company, 1987.

Lawrie, John. *You Can Lead*. New York: Amacom, 1985.

Lewis, James, Jr. *Excellent Organizations: How to Develop and Manage Them Using Theory Z*. New York: J. L. Wilkerson Publishing Co., 1985.

Luther, William M. *How to Develop a Business Plan in 15 Days*. New York: Amacom, 1987.

McLaughlin, Harold J. *Building Your Business Plan: A Step-By-Step Approach*. New York: Ronald Press, 1985.

McLuhan, Marshall. *Understanding Media: The Extension of Man*. New York: McGraw-Hill, 1965.

Migliore, R. Henry. *An MBO Approach to Long-Range Planning*. Englewood Cliffs, NJ: Prentice-Hall, Inc., 1983.

Mondy, R. Wayne, et al. *Management: Concepts and Practices, Ed. 3*. Boston: Allyn and Bacon, Inc. 1986.

Nelson, Robert B. *Decision Point: The Business Game That Lets You Make the Decisions*. New York: Amacom, 1987.

Oppenheimer, Matt and Young, Gerry A. *Computer-Assisted Business Plans*. Englewood Cliffs, NJ: Prentice-Hall, Inc., 1986.

Peters, Tom and Austin, Nancy. *A Passion for Excellence: The Leadership Difference*. New York: Random House, 1985.

Plachy, Roger. *When I Lead Why Don't They Follow?* Chicago: Bonus Books, 1986.

Porter, Michael E. *Competitive Advantage*. New York: Free Press, 1985.

Postman, N. *Amusing Ourselves to Death*. New York: Viking, 1985.

Raymond, H. Alan. *Management in the Third Wave*. Glenview, IL: Scott, Foresman and Co., 1986.

Ricketts, Martin. *The New Industrial Economics*. New York: St. Martin's Press, 1987.

Ritti, R. Richard and Funkhouser, G. Ray. *The Ropes to Skip and the Ropes to Know, Ed. 3*. New York: John Wiley & Sons, 1987.

Roth, William F., Jr. *Problem Solving for Managers*. New York: Praeger Special Studies, 1985.

Rothschild, William E. *How to Gain (and Maintain) the Competitive Advantage in Business*. New York: McGraw-Hill, 1984.

Sathe, Vijay. *Culture and Related Corporate Realities*. Homewood, IL: Richard D. Irwin, Inc., 1985.

Sawyer, George C. *Designing Strategy: A How-To Book for Managers*. New York: John Wiley & Sons, 1986.

Schatz, Kenneth and Schatz, Linda. *Managing By Influence*. Englewood Cliffs, NJ: Prentice-Hall, Inc., 1986.

Schein, Edgar H. *Organizational Culture and Leadership*. San Francisco:

Jossey-Bass Publishers, 1985.
Schmidt, William. *Media Center Management*. New York: Hastings House, 1980.
Scott, Bill and Soderberg, Sven. *The Art of Managing*. New York: John Wiley and Sons, 1985.
Souter, Gerald A. *The Disconnection: How to Interface Computers and Video*. White Plains, NY: Knowledge Industry Publications, Inc., 1988.
Starling, Grover. *The Changing Environment of Business, Ed. 2*. Boston, MA: Kent Pub. Co., 1984.
Stokes, Judith. *The Business of Nonbroadcast Television*. White Plains, NY: Knowledge Industry Publications, Inc., 1988.
Toffler, Alvin. *The Third Wave*. New York: Morrow & Co., 1980.
Van Deusen, Richard E. *Practical AV/Video Budgeting*. White Plains, NY: Knowledge Industry Publications, Inc., 1984.
Vos, Hubert D. *What Every Manager Needs to Know About Finance*. New York: Amacom, 1986.
Weiss, W. H. *Decision Making for First-Time Managers*. New York: Amacom, 1985.
Wershing, Stephen and Singer, Paul. *Computer Graphics and Animation for Corporate Video*. White Plains, NY: Knowledge Industry Publications, Inc., 1988.
Widner, Doug. *Teleguide: A Handbook on Video Teleconferencing*. Washington, DC: Public Service Satellite Consortium, 1986.
Williams, Frederick. *The Communications Revolution*. New York, NY: Mentor Books, 1983.
Winston, Brian. *Misunderstanding Media*. Cambridge, MA: Harvard University Press, 1986.

Appendix II: Trade and Professional Organizations

- **AAVT**
  Association of Audio-Visual Technicians
  PO Box 9716
  Denver, Co 80209
  303-698-1820
  Production technicians—graphic artists, photographers, etc.; equipment and maintenance repair persons. Also: trains technicians, offers placement services.

- **ABC**
  Association for Business Communication
  English Bldg.
  608 S. Wright St.
  University of Illinois
  Urbana, IL 61801
  217-333-1007
  College teachers of business communication and management consultants.

- **ACVL**
  Association of Cinema & Video Laboratories
  c/o Burton Stone
  Deluxe Laboratories, Inc.
  1377 N. Serrano Ave.
  Hollywood, CA 90027
  213-462-6171
  Motion picture laboratories supplying various services to motion picture producers, agencies, television and the theater.

- **AECT**
  Association for Educational Communications & Technology
  1126 16th St., NW
  Washington, DC 20036
  202-466-4780
  AV and instructional materials specialists, educational technologists, AV and TV production personnel and educators.

- **AES**
  Audio Engineering Society
  60 E. 42nd St., Room 2520
  New York, NY 10065
  212-661-8528
  Engineers, administrators and technicians who design or operate recording equipment for radio, TV, motion picture and recording studios.

- **AICP**
  Association of Independent Commercial Producers
  100 E. 42nd St., 16th Floor
  New York, NY 10017
  212-867-5720
  Independent producers of television commercials, associates supply editorial and payroll services, set designers, optical houses, labs.

- **AIGA**
  American Institute of Graphic Arts
  1059 Third Ave.
  New York, NY 10021
  212-752-0813
  Graphic artists involved in book design, illustrations, advertising, corporate graphics, promotion and exhibitions.

- **AIM**
  American Institute of Management
  45 Willard St.
  Quincy, MA 02169
  617-472-0277
  Executives interested in management efficiency and methods of appraising management performance.

- **AITV**
  Association of Independent TV Stations
  1200 18th St. NW, Suite 502
  Washington, DC 20036
  202-887-1970
  Commercial, independent television broadcasting stations not primarily affiliated with a national television network.

- **AMA**
  American Management Association
  135 W. 50th St.
  New York, NY 10020
  212-586-8100
  Managers in industry, commerce, government.

- **AMI**
  Association for Multi-Image
  8019 N. Himes Ave., Suite 901
  Tampa, FL 33614
  813-932-1692
  Educators, industrial trainers, media specialists.

- **ANA**
  Association of National Advertisers
  155 E. 44th St.
  New York, NY 10017
  212-697-5950
  National and regional advertisers; committees on AV and sales promotion.

- **ARSC**
  Association for Recorded Sound Collections
  c/o Phillip Rochlin
  PO Box 75082
  Washington, DC 20013
  703-591-6746
  Persons in the broadcasting and recording industries, librarians, curators, private collectors and audio archivists.

- **ASC**
  American Society of Cinematographers
  1782 N. Orange Dr.
  Hollywood, CA 90028
  213-876-5080
  Professional directors of motion picture and television photography; others affiliated with cinematography.

- **ASMP**
  American Society of Magazine Photographers
  205 Lexington Ave.
  New York, NY 10016
  212-889-9144
  Maintains and promotes high professional standards and ethics in photography; cultivates mutual understanding among professional photographers.

- **ASP**
  American Society of Photographers
  PO Box 52900
  Tulsa, OK 74152
  918-743-2122
  Photographers who have earned masters or other professional degrees in the field.

- **ASTD**
  American Society for Training and Development
  Box 1443
  1630 Duke St.
  Alexandria, VA 22313
  703-683-8100
  Persons engaged in training and development of business, industrial and government personnel.

- **ASTVC**
  American Society of TV Cameramen
  PO Box 296
  Sparkill, NY 10976
  914-359-5569
  Professional cameramen and persons in related jobs.

- **ATAS**
  Academy of Television Arts and Sciences
  3500 West Olive Ave., Suite 700
  Burbank, CA 91505
  818-953-7575
  Professionals in the television and film industry.

- **ATSC**
  Advanced TV Systems Committee
  1771 N St., NW
  Washington, DC 20036
  202-429-5345
  Members of the television and motion picture industry united to develop voluntary national standards in the area of advanced television systems.

- **AVC**
  Association of Visual Communicators
  900 Palm Ave., Suite B
  South Pasadena, CA 91030
  818-441-2274
  Media producers, managers, creative and technical people involved in film, videotape, filmstrips, slide films and multi-image audio and videodisc production.

- **AVMA (formerly IAVA)**
  Audio Visual Management Association
  PO Box 165887
  Irving, TX 75016
  Managers of AV departments of business and industrial firms.

- **AWRT**
  American Women in Radio & TV
  1101 Connecticut Ave., NW, Suite 700
  Washington, DC 20036
  202-429-5102
  Professionals in administrative, creative or executive positions in broadcasting and ad agencies, service organizations, government or charitable agencies devoted to radio and TV.

- **BDA**
  Broadcast Designers Association
  251 Kearny St., Suite 602
  San Francisco, CA 94108
  415-788-2324
  Designers, artists, art directors, illustrators, photographers, animators and other professionals in the television industry; educators and students; commercial and industrial companies that manufacture products related to design.

- **CCA**
  Cooperative Communicators Association
  2263 E. Bancroft
  Springfield, MO 65804
  417-882-1493
  Communicators in various disciplines provide assistance in writing, editing, photography, layout, speech writing, use of audiovisuals and communications management.

- **CCIA**
  Computer and Communications Industry Association
  666 11th St., NW, 6th floor
  Washington, DC 20001
  202-783-0070
  Mainframe vendors, software and service houses, leasing and maintenance companies, peripheral manufacturers.

- **CCM**
  Council of Communication Management
  PO Box 3970, Grand Central Post Office
  New York, NY 10163
  Individuals who work in communications with business and government organizations and who have responsibility for establishing policy and directing various media.

- **CINE**
  Council on International NonTheatrical Events
  1201 16th St., NW
  Washington, DC 20036
  202-785-1136
  Directors and individuals concerned with nontheatrical, television documentary, and short subject motion pictures and videotapes.

- **EAT**
  Experiments in Art & Technology
  49 East 68th St.
  New York, NY 10021
  212-285-1690
  Artists, engineers, scientists, composers, dancers and educators initiate projects based on collaboration between scientists and artists.

- **EFLA**
  Educational Film Library Association
  45 John St., Suite 301
  New York, NY 10010
  212-227-5599
  Educational institutions, commercial organizations and individuals interested in nontheatrical films. National clearinghouse for information about 16mm films.

- **EIA**
  Electronic Industries Association
  2001 I St., NW
  Washington, DC 20006
  202-457-4900
  Manufacturers of radio, TV, video systems, audio equipment and industrial and communications electronic products.

- **8mm Video Council**
  99 Park Ave.
  New York, NY 10016
  212-986-3978
  8mm videotape hardware, software and blank tape companies; video service and publications companies; accessory manufacturers and other trade associations.

- **Electronic VIP Club**
  c/o Sanford Levey
  Electronic Distributors, Inc.
  4900 N. Elston Ave.
  Chicago, IL 60630
  312-283-4800
  Recognizes men and women who have made significant contributions to the electronics industry.

- **FLN**
  Freelance Network
  PO Box 36838, Miracle Mile Station
  Los Angeles, CA 90036
  213-655-4476
  Freelance communications specialists including artists, photographers, writers, designers, editors, illustrators, advertisers and publishing consultants.

- **FSI**
  Freelance Syndicate, Inc.
  PO Box 1626
  Orem, UT 84057
  801-785-1300
  Freelancers of various types including writers, graphic artists, and photographers.

- **GAAEC**
  Graphic Arts Advertisers and Exhibitors Council
  c/o Garret W. Walker
  Baldwin Technology Corp.
  417 Shippan Ave.
  Stamford, CT 06902
  203-348-4400
  Advertising, marketing and trade show managers of manufacturers and suppliers that sell products to the graphic arts industry.

- **GAG**
  Graphic Artists Guild
  30 E. 20th St.
  New York, NY 10003
  212-777-7353
  National organization for commercial artists, illustrators, designers, etc.

- **GS**
  Goudy Society
  Horace Hart
  6219 Canadice Hill Rd.
  Springwater, NY 14560
  716-367-2851
  Individuals, firms and organizations interested in the understand-

ing of printed communications, promotes appreciation and practice of craftsmanship in typography and printing.

- **HESCA**
  Health Science Communications Association
  6105 Lindell Blvd.
  St. Louis, MO 63112
  314-725-4722
  Media managers, graphic artists, biomedical librarians, producers of programs, faculty members of health science schools, industry representatives.

- **HRTS**
  Hollywood Radio and Television Society
  5315 Laurel Canyon Blvd.,
  Suite 202
  North Hollywood, CA 91607
  818-769-4313
  Persons involved in radio, television, broadcasting and advertising, including programmer and commercial producers and radio and television networks and studios seeking to promote the broadcasting industry.

- **IABC**
  International Association of Business Communicators
  870 Market St., Suite 940
  San Francisco, CA 94102
  415-433-3400
  Communication managers, public relations directors, writers, editors, audiovisual specialists, and others in the public relations and communications fields who use a variety of media to communicate with internal and external audiences.

- **IAFS**
  International Animated Film Society
  5301 Laurel Canyon Blvd.
  Suite 250
  North Hollywood, CA 91607
  818-508-5224
  Exchange information on animated film and technique; support animated film societies; restores and repairs damaged cels.

- **IAS**
  International Audiovisual Society
  c/o Dr. Paul S. Flynn
  Western Carolina Univ.
  Cullowhee, NC 28723
  Individuals with academic degrees or equivalent in scientific or technical experience, with expertise in AV technology principles.

- **ICA**
  International Communications Association
  12750 Merit Drive, Suite 710 LB-89
  Dallas, TX 75251
  214-233-3889
  Representatives who are responsible for telecommunications services of major corporations and other organizations.

- **IDA**
  International Documentary Association
  8480 Beverly Blvd., Suite 140
  Los Angeles, CA 90048
  213-655-7089
  Individuals and organizations involved in nonfiction film and video.

- **IEEE**
  Institute of Electrical and Electronics Engineers
  345 E. 47th St.
  New York, NY 10017
  212-705-7900
  Engineers, scientists and students in electrical engineering, electronics and allied fields.

- **IGC**
  Institute for Graphic Communication
  375 Commonwealth Ave.
  Boston, MA 02115
  617-267-9425

Persons interested in graphic communications technologies and markets.

- **IIA**
  Information Industry Association
  555 New Jersey Ave., NW,
  Suite 800
  Washington, DC 20001
  202-639-8262
  Trade association of companies who are interested and involved in the business opportunities associated with the generation, distribution and use of information.

- **IICS**
  International Interactive Communications Society
  2120 Steiner St.
  San Francisco, CA 94115
  Individuals and organizations interested in interactive video involved in education, point-of-purchase retailing, and archival storage.

- **IIE**
  Institute of Industrial Engineers
  25 Technology Park/Atlanta
  Norcross, GA 30092
  404-449-0460
  Industrial engineers and students concerned with the design, improvement and installation of integrated systems of people, materials, equipment and energy.

- **IMPC**
  Independent Media Producers Council
  3150 Spring St.
  Fairfax, VA 22031
  703-273-7200
  Individuals and firms engaged in producing motion pictures, videotapes, slide shows and audio presentations for outside clients.

- **Intermedia**
  475 Riverside Dr., Room 670
  New York, NY 10115
  212-870-2376
  Internationally promotes literacy and literature programs, audiovisual production centers, radio and television programming and operations, publishing houses and bookstores, and workshops and training conferences.

- **International Communications Industry Association**
  3150 Spring St.
  Fairfax, VA 22031
  703-273-7200
  Dealers, manufacturers, producers and suppliers of AV products and materials.

- **IQ**
  International Quorum of Film and Video Producers
  PO Box 395
  Oakton, VA 22124
  703-648-0818
  Nontheatrical motion picture production companies specializing in films for industry, government and television.

- **IRTS**
  International Radio & TV Society
  420 Lexington Ave.
  New York, NY 10170
  212-867-6650
  Individuals in management, sales, or executive production in radio and TV broadcasting industries and their allied fields.

- **ISCET**
  International Society of Certified Electronics Technicians
  2708 W. Berry, Suite #8
  Ft. Worth, TX 76109
  817-921-9101
  Technicians who have been certified by the society, concerned with improving the effectiveness of industry education programs for technicians.

- **ISV**
  International Society of Videographers
  c/o American Society of TV Cameramen

Box 296
Washington St.
Sparkill, NY 10976
914-359-5569
Operations personnel from broadcasting, cable and television production houses; individual videographers and technicians; individuals associated with instructional television institutions.

- **ITA**
International Tape/Disc Association
10 Columbus Circle
New York, NY 10019
212-956-7110
Major manufacturers of home and institutional audio-videotape and disc equipment and supportive industries.

- **ITS**
International Teleproduction Society
990 Avenue of the Americas, Suite 21E
New York, NY 10018
212-629-3266
Professionals and others interested in making videotapes.

- **ITVA**
International Television Association
6311 N. O'Connor Rd., Suite 110-LB 51
Irving, TX 75039
214-869-1112
Persons engaged in communications needs analysis, scriptwriting, producing, directing, operations and management in nonbroadcast television.

- **IVLA**
International Visual Literacy Association
c/o Alice Walker
Virginia Polytechnic Institute and State University
Instructional Development Learning Resource Center
Old Security Bldg.
Blackburg, VA 24061
703-961-5879
Professionals in teaching (visual media, early learning), medicine, and television interested in methods of visual communication.

- **JCET**
Joint Council on Educational Telecommunications
c/o Corp. for Public Broadcasting
1111 16th St., NW
Washington, DC 20036
202-955-5278
Leading communications and educational organizations; coordinates education's interests in communications technology, policy and regulation.

- **MCEI**
Marketing Communications Executives International
c/o Dennis D'amico
2602 McKinney Ave.
Dallas, TX 75204
214-871-1016
Executives engaged in supervision, planning, execution or direction of marketing communications.

- **NAB**
National Association of Broadcasters
1771 N St., NW
Washington, DC 20036
202-429-5300
Radio and TV stations, associate producers of equipment and programs.

- **NAMAC**
National Alliance of Media Arts Centers
c/o Robert Haller
135 St. Paul's Ave.
Staten Island, NY 10301
718-727-5593
Media centers, museums, independent producers, filmmakers, video

artists, universities and professional service organizations that provide services for production, education, exhibition, preservation and distribution of video, film, audio and intermedia.

- **NAMW**
  National Association of Media Women
  1185 Niskey Lake Rd., SW
  Atlanta, GA 30331
  404-344-5862
  Women professionally engaged in mass communications.

- **NAPET**
  National Association of Photo Equipment Technicians
  3000 Picture Place
  Jackson, MI 49201
  517-788-8100
  Providers of camera repair services.

- **NAPM**
  National Association of Photographic Manufacturers
  600 Mamaroneck Ave.
  Harrison, NY 10528
  914-698-7603
  Manufacturers of photographic and other imaging equipment, supplies, films, chemicals.

- **NATAS**
  National Academy of TV Arts and Sciences
  110 W. 57th St.
  New York, NY 10019
  212-586-8424
  Persons actively engaged in TV performing, art direction, taping, tape editing, etc.

- **NAVD**
  National Association of Video Distributors
  1255 23rd St., NW
  Washington, DC 20037
  202-452-8100
  Wholesale distributors of home video software including discs and cassettes.

- **NCA**
  National Composition Association
  1730 N. Lynn St.
  Arlington, VA 22209
  703-841-8165
  Typesetting, word processing, computerized equipment users.

- **NCGA**
  National Computer Graphics Association
  2722 Merrilee Dr., Suite 200
  Fairfax, VA 22031
  703-698-9600
  Individuals and organizations who use, manufacture and sell computer graphics hardware and software.

- **NCTA**
  National Cable TV Association
  1724 Massachusetts Ave., NW
  Washington, DC 20036
  202-775-3550
  Cable TV systems, cable and equipment manufacturers, distributors, brokerage firms and financial institutions.

- **NCTI**
  National Cable TV Institute
  PO Box 27277
  Denver, CO 80227
  303-761-8554
  Provides technical educational material for the cable TV industry.

- **NFLPA**
  National Free Lance Photographers Association
  Ten S. Pine St.
  Doylestown, PA 18901
  215-348-5578
  Amateur and professional photographers. Maintains photographic file for industry members.

- **NVC**
  National Video Clearinghouse

100 Lafayette Dr.
Syosset, NY 11791
516-364-3686
Publisher of reference books detailing available video programming. Compiles statistics. Publishes The Video Source Book and the Videotape/Disc Guide to Home Entertainment.

- **NY/IABC**
New York/International Association of Business Communicators
PO Box 2025
Grand Central Station
New York, NY 10017
212-957-6661
Editors of company publications, communications specialists, managers of communications programs.

- **PFVEA**
Professional Film and Video Equipment Association
3211 South La Cienega Blvd., Suite C
Culver City, CA 90016
213-479-2549
Manufacturers, distributors, dealers and servicers of professional film and video equipment.

- **POPAI**
Point-of-Purchase Advertising Institute
Two Executive Dr.
Ft. Lee, NJ 07024
201-585-8400
Producers and suppliers of point-of-purchase advertising signs and displays. Retailers interested in use and effectiveness of point-of-purchase media are associate members.

- **PPofA**
Professional Photographers of America, Inc.
1090 Executive Way
Des Plaines, IL 60018
312-299-8161
Portrait, commercial and industrial photographers. Includes the American Photographic Artisan's Guild and the American Society of Photographers.

- **PRSA**
Public Relations Society of America
845 Third Ave.
New York, NY 10022
212-826-1750
Public relations practitioners in business and industry, counseling firms, trade and professional groups, government, education, health and welfare organizations.

- **PWP**
Professional Women Photographers
c/o Photographics Unltd.
17 West 17th St., #14
New York, NY 10011
212-255-9676
Women professional photographers and other interested individuals.

- **SALT**
Society for Applied Learning Technology
50 Culpeper St.
Warrenton, VA 22186
703-347-0055
Senior executives from military, academic and industrial organizations which design, manufacture, or use training technology including audiovisual instruction delivery devices.

- **SBCA**
Satellite Broadcasting and Communications Association
c/o Chuck Hewitt
300 North Washington St., Suite 310
Alexandria, VA 22314
703-549-6990
Equipment manufacturers of satellite earth stations, dealers and distributors; owners and operators.

- **SBE**
Society of Broadcast Engineers
7002 Graham Rd., Suite 118

Indianapolis, IN 46220
317-842-0836
Broadcast engineers, students and broadcasting professionals in closely allied fields.

- **SMPTE**
Society of Motion Picture and Television Engineers
595 W. Hartsdale Ave.
White Plains, NY 10607
914-761-1100
Professional engineers and technicians in motion picture, television and allied arts and sciences.

- **SPAR**
Society of Photographers and Artist Reps
1123 Broadway, #914
New York, NY 10010
212-924-6023
Persons who represent commercial artists and photographers to the graphic art world.

- **STC**
Society for Technical Communication
815 15th St., NW, Suite 506
Washington, DC 20005
202-737-0035
Educators, scientists, engineers, artists and others, professionally engaged in corporate technical communications.

- **TIA (formerly ITCA)**
Typographers International Association
2262 Hall Place, NW
Washington, DC 20007
202-965-3400
Typographic companies serving printers, publishers, advertising agencies, art studios, government, institutions and buyers of typographic composition and related products and services.

- **TMDA**
Training Media Distributors Association
25605 Cielo Ct.
Valencia, CA 91355
805-254-7224
Media producers and distributors interested in combating unauthorized copying of film.

- **Typophiles**
140 Lincoln Rd.
Brooklyn, NY 11225
718-462-2017
Designers, printers, book collectors, calligraphers, private press owners and others interested in the graphic arts.

- **VAPA**
Video Alliance for the Performing Arts
c/o Home Poupart
82 West 12th St.
New York, NY 10011
212-929-9107
Purpose is to assemble and distribute educational tapes, films and multi-visual presentations to educational TV, PBS, schools and civic groups.

- **VDC**
Video-Documentary Clearinghouse
Harbor Square, Suite 2201
700 Richards St.
Honolulu, HI 96813
808-523-2882
Video-documentary archive that works to improve the quality of higher education and organizational training by compiling a historical videotape archive of major contributors to important bodies of knowledge.

- **VIA**
Videotex Industry Association
1901 North Ft. Myer Dr., Suite 200
Rosslyn, VA 22209
703-522-0883
Firms engaged in manufacturing, designing, publishing, etc. videotex and teletext equipment or services in the United States.

- **VSDA**
  Video Software Dealers Association
  Three Eyes Drive, Suite 307
  Marlton, NJ 08053
  609-596-8500
  Retailers and wholesalers of videocassettes and videodiscs; studios and independent companies that produce video programming and manufacturers of video games, accessories, and services for video software.

- **WBT**
  Women in Broadcast Technology
  c/o Susan Elisabeth
  2435 Spaulding St.
  Berkeley, CA 94703
  415-642-1311
  Women employed in broadcast-related technology fields; media students.

- **WCA**
  World Communication Association
  c/o Ronald L. Applbaum
  Pan American University
  Edinburg, TX 78539
  512-381-2111
  Seeks to promote academic phases of communications, particularly radio, television, film, oral, written and electronic communication.

- **WGA**
  Writers Guild of America, West
  8955 Beverly Blvd.
  Los Angeles, CA 90048
  213-550-1000
  Labor union for writers in motion pictures, TV and radio.

- **WGAE**
  Writers Guild of America, East
  555 W. 57th St.
  New York, NY 10019
  212-245-6180

- **WIC**
  Women in Cable
  c/o P.M. Haeger & Associates
  500 N. Michigan Ave., Suite 1400
  Chicago, IL 60611
  312-661-1700
  Individuals in cable television and related industries and disciplines.

- **WICI**
  Women in Communications, Inc.
  PO Box 9561
  Austin, TX 78766
  512-346-9875
  Women in journalism and communications.

- **WIF**
  Women in Film
  6464 Sunset Blvd.
  Hollywood, CA 90028
  213-463-6040
  Purpose is to support women in the film and television industries and to serve as a network for information on qualified women in the entertainment fields.

- **WIP**
  Women In Production
  211 East 43rd St.
  New York, NY 10017
  212-867-4194
  Persons involved in all phases of graphic arts.

- **YBPC**
  Young Black Programmers Coalition
  PO Box 11243
  Jackson, MS 39213
  601-634-5775
  Black professionals in communications, broadcasting and music industries; provides training and technical assistance.

Appendix III: Trade and Professional Periodicals

- **American Cinematographer**
American Society of Cinematographers Holding Corp.
Box 2230
Hollywood, CA 90078

- **American Photographer**
CBS Magazines
1515 Broadway
New York, NY 10036

- **Annual of Advertising and Editorial Art & Design**
Reinhold Publishing Corp.
600 Summer St.
Stamford, CT 06901

- **Audio**
CBS Magazines
1515 Broadway
New York, NY 10036

- **Audio/Video**
Society of Audio/Video Consultants
PO Box 660
Beverly Hills, CA 90213

- **Audio/Video ICs D.A.T.A. Book**
D.A.T.A., Inc.
9889 Willow Creek Rd.
San Diego, CA 92131

- **AudioVideo International**
Dempa Publications, Inc.
380 Madison Ave.
New York, NY 10017

- **Audio Video Market Place**
Hilary House Publishers
1033 Channel Dr.
Hewlett, NY 11557

- **Audio Visual Communications**
Media Horizons, Inc.
50 W. 23rd St.
New York, NY 10010

- **Audio-Visual Equipment**
Media Horizons, Inc.
50 W. 23rd St.
New York, NY 10010

- **A V Video**
Montage Publishing
25550 Hawthorne Blvd., Suite 314
Torrance, CA 90505

- **Back Stage**
Back Stage Publications
330 W. 42nd St.
New York, NY 10036

- **Board Report for Graphic Artists**
Board Report Publishing Co., Inc.
Box 1561
Harrisburg, PA 17105

- **Broadcast Engineering**
Intertec Publishing Corp.
Box 12901
Overland Park, KS 66212

- **Broadcast Management/Engineering**
Broadcast Management/Engineering
295 Madison Ave.
New York, NY 10017

- **Broadcasting**
Broadcasting Publications, Inc.
1735 DeSales St., NW
Washington, DC 20036

- **Business & the Media**
Media Institute
3017 M St., NW
Washington, DC 20007

- **Business Marketing**
Crain Communications
220 E. 42nd St.
New York, NY 10017

- **Channels**
  Public Relations Publishing Co.
  Box 600, Dudley House
  Exeter, NH 03833

- **C.I.R. Communication Industry Report**
  International Communication Industries Association
  3150 Spring St.
  Fairfax, VA 22031

- **Communications News**
  Harcourt Brace Jovanovich
  402 W. Liberty
  Wheaton, IL 60187

- **The Communicator**
  IFPA Film & Video Communicators
  3518 Cahuenga Blvd. W.
  Suite 313
  Hollywood, CA 90068

- **Communications World**
  International Association of Business Communications
  870 Market St., Suite 940
  San Francisco, CA 94102

- **Corporate Communications Digest**
  Stein Printing Co.
  2161 Monroe Dr., NE
  Atlanta, GA 30324

- **Corporate Times**
  EML Publications
  1214 Oakmead Pkwy., Bldg. A
  Sunnyvale, CA 94086

- **Creative Forum**
  Direct Marketing Creative Guild
  516 Fifth Ave.
  New York, NY 10036

- **DB-The Sound Engineering Magazine**
  Sagamore Publishing Co., Inc.
  1120 Old Country Rd.
  Plainview, NY 11803

- **Display & Imaging Technology**
  Gordon Breach Science Publishers
  Box 786 Cooper Station
  New York, NY 10276

- **ECTJ Educational Communicators and Technicians Journal**
  Association For Educational Communicators
  1126 16th St., NW
  Washington, DC 20036

- **Electronic News**
  Fairchild Publications
  7 E. 12th St.
  New York, NY 10003

- **Electronics Industries Association, Executive Report**
  Electronics Industries Association
  2001 I St., NW
  Washington, DC 20006

- **Electronic's Week**
  McGraw Hill
  1221 Avenue of the Americas, Suite 4360
  New York, NY 10020

- **Electronics Weekly**
  Business Press International
  205 E. 42nd St.
  New York, NY 10017

- **ETV Newsletter**
  C.S. Tepfer Publishing Co.
  51 Sugar Hollow Rd.
  Danbury, CT 06810

- **The Executive Manager**
  American Association of Industrial Management
  2500 Office Center
  Maryland Rd.
  Willow Grove, PA 19090

- **Filmmakers Film and Video Monthly**
  Suncraft International, Inc.
  PO Box 115
  Ward Hill, MA 01801

- **Functional Photography**
  PTN Publishing Corp.
  210 Crossways Park Dr.
  Woodbury, NY 11797

- **Graphics Today**
  Communication Channels, Inc.
  6255 Barfield Road
  Atlanta, GA 30328

- **Graphics; USA**
  120 E. 56th St.
  New York, NY 10022

- **Home Video Marketplace**
  Spin Pubs.
  1965 Broadway
  New York, NY 10023

- **IEEE Transactions on Professional Communication**
  IEEE
  345 E. 47th St.
  New York, NY 10017

- **Illustrators 27**
  Madison Square Press
  10 E. 23rd St.
  New York, NY 10010

- **Independent Advertising & Marketing Services**
  Executive Communications, Inc.
  919 Third Ave., 17th Floor
  New York, NY 10022

- **Industrial Marketing Management**
  Elsevier Science Publishing Co.
  52 Vanderbilt Ave.
  New York, NY 10017

- **Industrial Photography**
  Media Horizons, Inc.
  50 W. 23rd St.
  New York, NY 10010

- **Innovation, Journal of Industrial Design**
  Industrial Designers Society of America
  6802 Popular Place
  McLean, VA 22101

- **International Documentary**
  International Documentary Foundation
  8480 Beverly Blvd., Suite 140
  Los Angeles, CA 90048

- **International Photographer**
  International Alliance of Theatrical State Employees and Moving Picture Machine Operators
  7715 Sunset Blvd., Suite 150
  Hollywood, CA 90046

- **International Technical Communications Conference Proceedings**
  Society for Technical Communications Distributors
  PO Box 28130
  San Diego, CA 92128

- **Journal of Biological Photography**
  Biological Photographic Association, Inc.
  115 Stoneridge Dr.
  Chapel Hill, NC 27514

- **Journal of Broadcasting**
  Broadcast Education Association
  1771 N St., NW
  Washington, DC 20036

- **Journal of Business Communications**
  Association for Business Communications
  English Bldg.
  608 S. Wright St.
  South Urbana, IL 61801

- **Journal of Communication**
  Annenberg School of Communications
  3620 Walnut St.
  Philadelphia, PA 19104

- **Journal of Educational Technology Systems**
  Baywood Publishing Co., Inc.
  120 Marine St., Box D
  Farmingdale, NY 11735

- **Journal of Micrographics**
  National Micrographics Association
  1100 Wayne Ave.
  Silver Springs, MD 20910

- **Journal of Technical Writing and Communication**
  Baywood Publishing Co.
  120 Marine St., Box D
  Farmingdale, NY 11735

- **Knowledge: Creation, Diffusion, Utilization**
  Sage Publications, Inc.
  275 S. Beverly Dr.
  Beverly Hills, CA 90212

- **Marketing Communications**
  Media Horizons, Inc.
  50 W. 23rd St.
  New York, NY 10010

- **Marketing & Media Decisions**
  Marketing & Media Decisions
  1140 Avenue of the Americas
  New York, NY 10036

- **Media History Digest**
  Media History Digest, Corp.
  11 W. 19th St.
  New York, NY 10011

- **Media Industry Newsletter**
  MIN Publishing Co.
  145 E. 49th St.
  New York, NY 10017

- **Media Mind**
  International Audiovisual Society
  PO Box 2
  Cullowhee, NC 28723

- **Media Science Newsletter**
  New Electronic Media Science, Inc.
  250 Mercer St., #1408B
  New York, NY 10012

- **Millimeter**
  Penton Pub.
  1111 Chester Ave.
  Cleveland, OH 44114

- **Modern Photography**
  ABC Consumer Magazines, Inc.
  825 7th Ave.
  New York, NY 10019

- **M.P.E. Audio-Visual Source Directory**
  Motion Picture Enterprises Publications
  PO Box 276
  Tarrytown, NY 10591

- **Multi-Images**
  Association for Multi-Image Intl.
  8019 N. Himes Ave., Suite 401
  Tampa, FL 33614

- **NAVA News**
  National Audio-Visual Association
  3150 Spring St.
  Fairfax, VA 22031

- **Optical Engineering**
  Society of Photo-Optical Instrumentation Engineers
  1022 19th St.
  Box 10
  Bellingham, WA 98225

- **Photogrammetric Engineering and Remote Sensing**
  American Society of Photogrammetry
  210 Little Falls St.
  Falls Church, VA 22046

- **Photographic Trade News**
  PTN Publishing Corp.
  210 Crossways Park Dr.
  Woodbury, NY 11797

- **Popular Photography**
  CBS Magazines
  1515 Broadway
  New York, NY 10036

- **Print**
  RC Publications Inc.
  355 Lexington Ave.
  New York, NY 10017

- **Printing Impressions**
  North American Publishing Co.
  401 N. Broad St.
  Philadelphia, PA 19107

- **Printing News**
  Printing News, Inc.
  468 Park Ave. S.
  New York, NY 10016

- **Professional Photographer**
  Professional Photographers of America, Inc.
  1090 Executive Way
  Des Plaines, IL 60018

- **Public Relations Journal**
  Public Relations Society of America
  845 Third Ave.
  New York, NY 10022

- **Publishers Weekly**
  Hilary House Pub.
  1033 Channel Dr.
  Hewlett, NY 11557

- **The Rangefinder**
  Rangefinder Publishing Co.
  1312 Lincoln Blvd.
  Box 1703
  Santa Monica, CA 90406

- **SMPTE Journal**
  Society of Motion Picture & TV Engineers
  862 Scarsdale Ave.
  Scarsdale, NY 10583

- **Satellite Communications**
  Cardiff Publishing Corp.
  6530 S. Yosemite
  Englewood, CO 80111

- **Satellite News**
  Phillips Publishing Inc.
  7811 Montrose Rd.
  Potomac, MD 20854

- **Science and Technology**
  Univelt Inc.
  Box 28130
  San Diego, CA 92128

- **Sightlines**
  EFLA-Educational Film Library Association
  45 John St., #301
  New York, NY 10038

- **Studio Photography**
  PTN Publishing Corp.
  210 Crossways Park Dr.
  Woodbury, NY 11797

- **Technical Photography**
  PTN Publishing Corp.
  210 Crossways Park Dr.
  Woodbury, NY 11797

- **TechTrends**
  Association for Educational Communications
  1126 16th St, NW
  Washington, DC 20036

- **Television/Radio Age**
  Television Editorial Corp.
  1270 Ave. of the Americas
  New York, NY 10020

- **Training**
  Lakewood Pubs., Inc.
  50 S. 9th St.
  Minneapolis, MN 55402

- **Training and Development Journal**
  ASTD-American Society for Training and Development
  600 Maryland Ave., SW
  Ste. 305
  Washington, DC 20024

- **U & L C**
  International Typeface Corp.
  2 Hammarskjold Plaza
  New York, NY 10017

- **Video Manager**
  Montage Publishing, Inc.
  25550 Hawthorne Blvd., Suite 314
  Torrance, CA 90505

- **Video Systems**
  Intertec Publishing Corp.
  9221 Quivira Rd.
  PO Box 12901
  Overland Park, KS 66212

- **Video Trade News**
  C.S. Tepfer Publishing Co.
  51 Sugar Hollow Rd.
  Danbury, CT 06810

- **Videography**
  P.S.N. Publications, Inc.
  2 Park Ave.
  New York, NY 10016

- **View Magazine**
  Macro Communications
  80 Fifth Ave.
  New York, NY 10011

- **Women in Communications Newsletter**
  Women in Communications, Inc.
  3724 Executive Center Dr.
  Austin, TX 78731

Appendix IV: Competitions

AMI Awards Festival and Competition
Association for Multi-Image
8019 North Himes Ave.
Suite 401
Tampa, FL 33614
813-932-1692
Date: August
Entry by May
Media: 2 projectors, 3-5 projectors, 6-8 projectors, 9-14 projectors, 15 or more projectors, video and video walls.
Categories: Commerical, noncommercial.

Intercom Industrial Film and Videotape Competition
415 North Dearborn St.
Chicago, IL 60610
312-644-3400
Date: April
Festival in September
Media: Videotape (3/4"), film (35mm, 16mm).
Categories: Business/industrial, training, health/medicine/safety.

Cine Golden Eagle Film Awards
Council on International Nontheatrical Events
1201 Sixteenth St., NW
Washington, DC 20036
202-785-1136
Date: May/October
Entry by February 1 or August 1
Media: Film (35mm, 16mm, super 8mm, 8mm) and videotape
Categories: Industry/commerce, technology, medicine/dental, entertainment, art, history, children's programming, et al.

Columbus International Film Festival
Columbus Film Council
1229 W. Third Ave.
Columbus, OH 43212
614-291-2149
Date: October
Entry by July
Media: Film and videotape
Categories: Business/industry, economics, employee relations, fund raising, industrial safety, manufacturing/technical, public realtions, personnel/sales training, sales promotion, education, social studies, religion, travel.

Industrial Photography Photo Contest
Industrial Photography Magazine
210 Crossways Park Dr.
Woodbury, NY 11797
516-496-8000
Date: June
Entry by February
Media: Prints, slides
Category: Open to industrial photographers.

ITVA International Videotape Competition
International Television Ass'n.
6311 N. O'Connor Rd., LB-51
Irving, TX 75039
Date: June
Entry: December
Media: Videotape
Categories: Training, information, sales/marketing, public service/public relations, company news, health and medicine, environment, education.

International Film and TV Festival of New York
International F.T.F. Corporation
5 W. 37th St.
New York, NY 10018
914-238-4481
Date: January
Entry: August
Media: Multimedia/mixed media, filmstrips/slide program, film (16mm), videotape (2" or cassette).
Categories: Corporate image, public relations, services training, manufacturing, product presentation.

Public Relations Film Festival
Public Relations Society of America
33 Irving Place
New York, NY 10003
212-995-2230
Date: November
Entry by March
Media: Film (16mm), videotape (3/4" cassette)
Categories: Corporate identity, institutional identity, community relations, internal communications, public education, product exposure.

San Francisco International Film Festival
1560 Fillmore St.
San Francisco, CA 94115
415-567-4641
Date: March
Entry by December
Media: Film (35mm, 16mm) and videotape
Categories: Nontheatrical, industry/business/government training, medical health, technical/scientific.

U.S. Industrial Film Festival
United States Festivals Association
841 North Addison Ave.
Elmhurst, IL 60126
312-834-7773
Date: April
Entry by March 1
Media: 16mm, videotape and videodisc, filmstrip, 35mm slides
Categories: In-plant, government, advertising/sales, promotional, recruiting, employee communications, industrial/technical, medicine/health.

International Monitor Awards
International Teleproduction Society
990 Avenue of the Americas, Suite 21E
New York, NY 10018
212-629-3266
Date: September
Entry: February
Media: Videotape
Categories: Broadcast and non-broadcast.

Telly Awards
4100 Executive Park Dr.
Cincinnati, OH 45241
513-421-1938
Date: February
Entry: December
Categories: Safety, training, sales, public relations, instructional, sports, how to, direct marketing.

Index

About the Author

Eugene Marlow has been involved with media for over 25 years—as a stage manager, radio producer and announcer, author, scriptwriter, musician and composer, stage, film and television producer/director and educator.

He is the author and contributing author to six books on communications media and has published more than 80 articles on broadcast television programming and video technologies in the United States and Europe.

From 1972 to 1982, Mr. Marlow held executive positions with Citibank, Prudential Insurance and Union Carbide Corporation. As head of video communications with Union Carbide, he supervised the design, construction and equipping of a $3 million production complex and a staff of 15 producers and technicians. In 1982, his department received "Department of the Year" honors from the Information Film Producers Association.

During this time, Mr. Marlow founded and chaired the New York chapter of the International Television Association and later served as its National Director of Professional Development. He taught communications at Fordham University's Graduate School of Public Communications from 1977 to 1979. He now teaches electronic journalism and business communications at Bernard M. Baruch College (City University of New York).

Prior to his entry into electronic media, Mr. Marlow was an award-winning United States Air Historian during the Vietnam War and a news editor for a mass merchandising trade journal.

Mr. Marlow is founder/president of Media Enterprises Inc, a New York-based company involved in a variety of electronic media production activities. Some of the company's clients include: *ADWEEK* Magazine, American Television & Communications, Armour Pharmaceutical, Bantam, CBS, Dupont, The Equitable, General Foods, Manhattan Cable

Television, New York Power Authority, Prudential-Bache Securities and Squibb.

Mr. Marlow has received more than 40 awards for programming excellence from a variety of national and international organizations. He also has extensive experience in the performing arts.

During 1987 to 1988 Mr. Marlow was the producer and host of ''Ad-News from *ADWEEK* Magazine'' which aired daily on WNCN-FM, New York. He also produced the national radio commercials for the 1988 JVC Jazz Festival.

Mr. Marlow has served as a blue-ribbon judge for children's and informational programming for the National Academy of Television Arts & Sciences and the National Academy of Cable Programming. He has also served as a judge for the Monitor Awards, the International Television Association, the International Film & Television Festival of New York and the McGraw-Hill Book Company Professional Recognition Awards for Advertising.

Mr. Marlow completed a PhD in media studies at New York University in 1988 and received an MBA from Golden Gate College in 1972. He is a member of the National Academy of Television Arts & Sciences (New York chapter) and the National Academy of Cable Programming.

Mr. Marlow lives in New York City with his wife, Judy, twin (fraternal) sons, Jonathan David and Samuel Josef, and five cats.